P9-CCC-136

TERPENOIDS: STRUCTURE, BIOGENESIS, AND DISTRIBUTION

Recent Advances in Phytochemistry

Volume 6

CONTRIBUTORS

Robert P. Adams
Pierre Crabbé
Rodney Croteau
Robert H. Flake
T. A. Geissman
T. W. Goodwin
Robert S. Irving
W. David Loomis
A. Ortega
Judith Polonsky
P. B. Reichardt
J. Romo
A. Romo de Vivar
A. I. Scott
M. B. Slaytor
J. G. Sweeny
B. L. Turner
Ernst Von Rudloff

TERPENOIDS: STRUCTURE, BIOGENESIS, AND DISTRIBUTION

Recent Advances in Phytochemistry

Volume 6

Edited by

V. C. RUNECKLES and T. J. MABRY

Department of Plant Science
University of British Columbia
Vancouver, British Columbia, Canada

Department of Botany
University of Texas
Austin, Texas

ACADEMIC PRESS New York and London 1973

ACADEMIC PRESS, INC.
111 Fifth Avenue, New York, New York 10003

United Kingdom Edition published by
ACADEMIC PRESS, INC. (LONDON) LTD.
24/28 Oval Road, London NW1

LIBRARY OF CONGRESS CATALOG CARD NUMBER: 77-182656

PRINTED IN THE UNITED STATES OF AMERICA

CONTENTS

The Biogenesis of Sesquiterpene Lactones of the Compositae

T. A. Geissman

Recent Developments in the Biosynthesis of Plant Triterpenes

T. W. Goodwin

Mechanisms of Indole Alkaloid Biosynthesis. Recognition of Intermediacy and Sequence by Short-Term Incubation

A. I. Scott, P. B. Reichardt, M. B. Slaytor, and J. G. Sweeny

Biochemistry and Physiology of Lower Terpenoids

W. David Loomis and Rodney Croteau

Genetic and Biosynthetic Relationships of Monoterpenes

Robert S. Irving and Robert P. Adams

Confirmation of a Clinal Pattern of Chemical Differentiation in *Juniperus virginiana* from Terpenoid Data Obtained in Successive Years

Robert H. Flake, Ernst Von Rudloff, and B. L. Turner

LIST OF CONTRIBUTORS

Number in parentheses indicate the pages on which the authors' contributions begin.

ROBERT P. ADAMS (187), Department of Botany and Plant Pathology, Colorado State University, Fort Collins, Colorado

PIERRE CRABBÉ (1), Universidad Nacional Autónoma de México and Research Laboratories Syntex, S. A., México

RODNEY CROTEAU (147), Department of Biochemistry and Biophysics, Oregon State University, Corvallis, Oregon

ROBERT H. FLAKE (215), Electrical Engineering Department, University of Texas, Austin, Texas

T. A. GEISSMAN (65), Department of Chemistry, University of California at Los Angeles, Los Angeles, California

T. W. GOODWIN (97), Department of Biochemistry, University of Liverpool, Liverpool, England

ROBERT S. IRVING (187),* Botany Department, University of Montana, Missoula, Montana

W. DAVID LOOMIS (147), Department of Biochemistry and Biophysics, Oregon State University, Corvallis, Oregon

A. ORTEGA (21), Instituto de Quimica de la Universidad Nacional Autónoma de México, México D. F., México

JUDITH POLONSKY (31), Institut de Chimie des Substances Naturelles, C. N. R. S., Gif-sur-Yvette, France

* Present address: Biology Department, Louisiana State University at New Orleans, New Orleans, Louisiana.

P. B. Reichardt* (117), Sterling Chemistry Laboratory, Yale University, New Haven, Connecticut

J. Romo (21), Instituto de Quimica de la Universidad Nacional Autónoma de México, México D. F., México

A. Romo de Vivar (21), Instituto de Quimica de la Universidad Nacional Autónoma de México, México, D. F., México

A. I. Scott (117), Sterling Chemistry Laboratory, Yale University, New Haven, Connecticut

M. B. Slaytor† (117), Sterling Chemistry Laboratory, Yale University, New Haven, Connecticut

J. G. Sweeny‡ (117), Sterling Chemistry Laboratory, Yale University, New Haven, Connecticut

B. L. Turner (215), Botany Department, University of Texas, Austin, Texas

Ernst Von Rudloff (215), Prairie Research Laboratory, National Research Council of Canada, Saskatoon, Canada

* Present address: Department of Chemistry, University of Alaska, Fairbanks, Alaska.

† Present address: Department of Biochemistry, The University of Sydney, Sydney, Australia.

‡ Present address: Organic Chemistry Laboratories, University of Technology, Loughborough, Leicestershire, England.

PREFACE

The present volume in the serial publication *Recent Advances in Phytochemistry* is derived from the Symposium of the Phytochemical Society of North America, held at the Instituto Technologico y de Estudios Superiores de Monterrey, Monterrey, Mexico, October 6–8, 1971. The symposium focused on advances in the chemistry and biochemistry of terpenoids, and the use of information regarding the occurrence of such compounds in genetics and population ecology.

No volume this size can attempt a comprehensive coverage of the whole field of terpenoid chemistry, biogenesis, and distribution. Nevertheless, the present volume presents a series of reviews of recent developments in a number of specialized areas of terpenoid research, commencing in the first article, by Pierre Crabbé, with a review of applications of physical techniques to structural and stereochemical problems. The second and third articles (J. Romo, A. Romo de Vivar, and A. Ortega; Judith Polonsky), respectively, describe the structural elucidations of new sesquiterpenes isolated from the Compositae and the chemistry and biosynthesis of the triterpenoid-derived quassinoid bitter principles which occur in the Simaroubaceae.

Biogenesis and metabolism are the themes of the next four articles. T. A. Geissman presents a view of sesquiterpene lactone biogenesis in the Compositae, while T. W. Goodwin focuses on recent developments in our knowledge of triterpene biosynthesis. In the sixth article (A. I. Scott, P. B. Reichardt, M. B. Slaytor, and J. G. Sweeny), the emphasis is upon indole alkaloid biosynthesis, with a description of the sorts of experimental approaches which may be employed for the elucidation of the biosynthetic pathways which lead to such complex metabolites. Terpenoid biosynthesis and metabolic turnover are dealt with by W. David Loomis and Rodney

Croteau, together with comments on the physiological function and distribution of these compounds within plants.

A detailed description of the genetic control of monoterpene biosynthesis in *Hedeoma* is covered by Robert S. Irving and Robert P. Adams, while the final article by Robert H. Flake, Ernst Von Rudloff, and B. L. Turner describes the use of terpene data in establishing clinal patterns of chemical differentiation in *Juniperus*.

The diversity of topics covered in this volume reflects the complexity of research being conducted in the terpenoid field.

As our knowledge of terpenoids continues to grow, we feel it is essential to have up-to-date reviews such as those presented here. We are especially grateful to the contributors of these articles and for the individual efforts they made to prepare and submit their manuscripts with the minimum of delay. Our thanks also go to Mrs. Shari Gris and Miss Diane Green, for secretarial work, and to the staff of Academic Press for their excellent cooperation in converting the symposium from manuscript to final book form.

V. C. Runeckles
T. J. Mabry

APPLICATIONS OF PHYSICAL METHODS TO SOME STRUCTURAL AND STEREOCHEMICAL PROBLEMS IN TERPENES AND STEROIDS

PIERRE CRABBÉ

Universidad Nacional Autónoma de México and Research Laboratories, Syntex, S. A., México

Introduction

Presently, the organic chemist has at his disposal a powerful and sophisticated armamentarium to solve structural and stereochemical problems. In addition to the time-honored classical infrared (IR) and ultraviolet (UV) techniques, more modern tools are now available, such as nuclear magnetic resonance (NMR) spectroscopy, gas–liquid chromatography (GLC), and mass spectrometry (MS). Combined with X-ray crystallography, these methods have considerably facilitated the work of the research scientist. Moreover, optical rotatory dispersion (ORD) and circular dichroism (CD) provide valuable additional information on the stereochemistry of complex molecular systems.

The first part of this report will mention some work undertaken at the

Facultad de Química de la Universidad Nacional Autónoma de México. Some years ago, the reaction of benzoyl chloride with *d*-pulegone (**1**) was investigated (Crabbé, 1954). Structures **2** and **3** were proposed for the compounds that were obtained, along with an unknown yellow crystalline substance. However, this work antedated modern techniques of NMR and MS spectroscopy. In addition, at that time there was no IR instrument available in our laboratory.

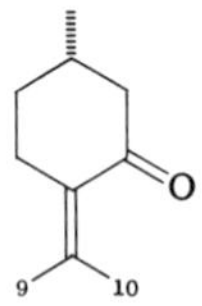

9 10 O

1

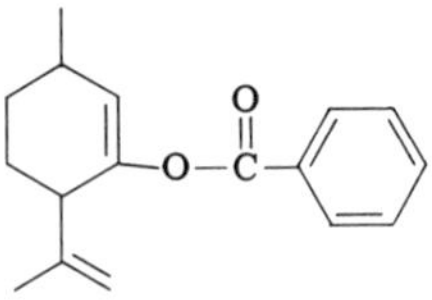

O-C(=O)

2

O-C(=O) C(=O)

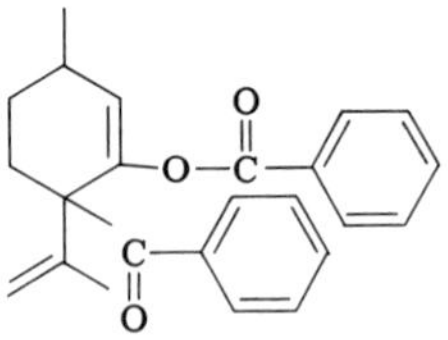

3

Recently, this reaction has been reexamined (Crabbé *et al.*, 1972a), and we shall now discuss the new structures assigned to these compounds. Identification of the reaction products was achieved by a combination of elemental analysis and NMR, MS, IR, and UV spectroscopy. In some cases the proposed structures were confirmed by chemical conversion into known compounds or into newly synthesized reference substances of assured constitution and stereochemistry.

Reexamination of Products of Benzoyl Chloride Reaction with *d*-Pulegone

d-Pulegone (**1**) was allowed to react with benzoyl chloride in benzene solution in the presence of sodium *t*-amylate (Crabbé, 1954; Crabbé *et al.*, 1972a). In addition to unreacted pulegone and *t*-amyl benzoate, an oily enol ester was isolated. The elemental analytical data and the molecular ion (m/e 256, M^+) support either structure **2** or **4**. The UV properties (λ_{max} 231 nm; log ϵ 4.23) indicate the summation of a diene and a benzoate chromophore, the latter being confirmed by the IR band at 1730 cm^{-1}. The NMR presents a signal at 1.83 ppm integrating for the 3 protons of the vinylic methyl at C–9. In addition, the only olefinic protons are the two C–10 methylene protons, which appear at 4.87 ppm, thus excluding structure **2**, previously assigned. Structure **4** was further confirmed by chemical transformations (Crabbé *et al.*, 1972a).

The second substance isolated from the reaction mixture is a crystalline material to which the β-ketoester structure **5** is being proposed, instead of the isomeric formula **3**. The MS shows a molecular ion at m/e 360 (M^+).

The UV spectrum (λ_{max} 237, 284, 308 nm; log ϵ 4.46, 3.79, 3.80) supports the extended conjugated system present in the molecule **5**. The NMR spectrum shows the secondary methyl at C–1 which appears as a doublet at 1.08 ppm (J 7 Hz). Moreover, two nonequivalent vinylic methyl signals are observed at 1.73 (C–9) and 1.76 ppm (C–10), respectively, in addition to ten aromatic protons. No vinylic hydrogens could be detected, in accord with structure **5**.

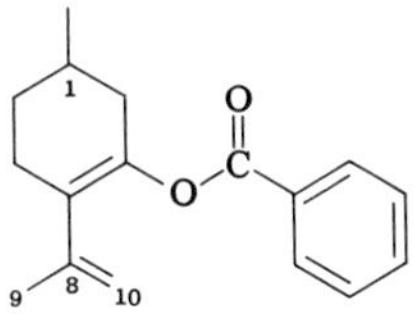

4 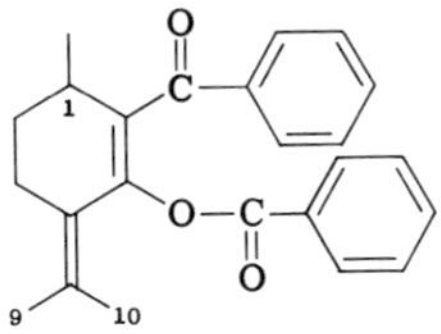 **5**

The reactions that the *β*-keto ester **5** undergoes with hydroxylamine are of a sophisticated nature, dictated by the complex functionality and the stereochemical sensitivity of the molecule to rather mild conditions. Thus treatment of **5** with hydroxylamine gives a substance whose MS shows a molecular ion *m/e* 286 (M^+), compatible with either formula **6** or **7**. The absence of a vinylic methyl at C–8 excludes structure **6**, and the appearance of a *gem*-dimethyl group at 1.50 and 1.65 ppm, supports the hypothesis that cyclization has taken place at C–8. The substituted isoxazoline structure **7** is further confirmed by the hydroxyl signal at 10.9 ppm,which integrates for one hydroxyl only, and is readily exchanged with deuterowater. This derivative **7** presumably arises via initial formation of the benzoyl oxime, followed by hydrolysis of the enol ester linkage. Subsequent reaction of the conjugated carbonyl with a second molecule of hydroxylamine then affords **7** by intramolecular cyclization.

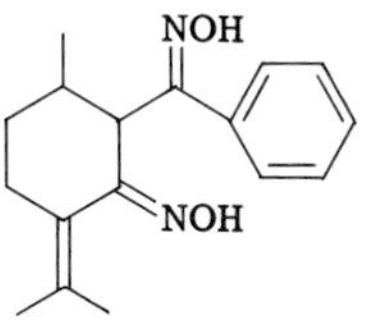

6 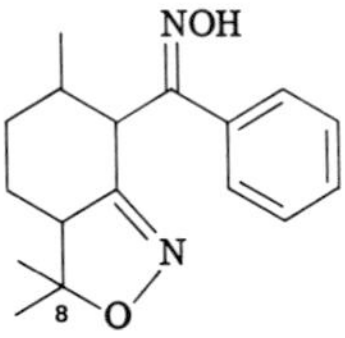 **7**

A third reaction product of **1** with benzoyl chloride is a crystalline material appearing as long yellow needles, shown to be identical with the unknown substance isolated earlier (Crabbé, 1954). The empirical formula $C_{24}H_{22}O_2$ deduced from the MS (m/e 342, M^+), was confirmed by elemental analysis. The IR spectrum presents a strong band at 1670 cm^{-1} indicative of a conjugated carbonyl and absorptions at 1620 and 690 cm^{-1} corresponding to the aromatic chromophore, in addition to another intense band at 1525 cm^{-1}, attributed to a 1,4-pyran ring (Nakanishi, 1964). Furthermore, the UV spectrum with its typical three absorption bands (λ_{max} 233, 295, 408 nm; log ϵ 4.18, 4.19, 4.50), indicates that a highly conjugated chromophore is present in the molecule.

Figure 1 reproduces the NMR of this substance which shows one doublet at 1.1 ppm (J 7 Hz) corresponding to the secondary methyl. Complex signals, integrating for seven alicyclic protons appear between 1.5 and 2.7 ppm. At 6.19 ppm there is a vinylic H signal corresponding to one proton. Multiplets integrating for ten aromatic H, appear in the 7.45–7.93 ppm region. Finally, a 1-proton signal appears at 9.0 ppm. Initially, this was assigned to an aldehydic proton. However, none of these protons exchanged with D_2O at room temperature, and all attempts to chemically identify an aldehyde group failed.

Hence, the 5-methylcyclohexanone-2′,6′-diphenylpyranylidene structure (**8**) was first assigned to this yellow product, although it was realized that the isomeric alternative (**9**) would also account for the physical and chemical properties.

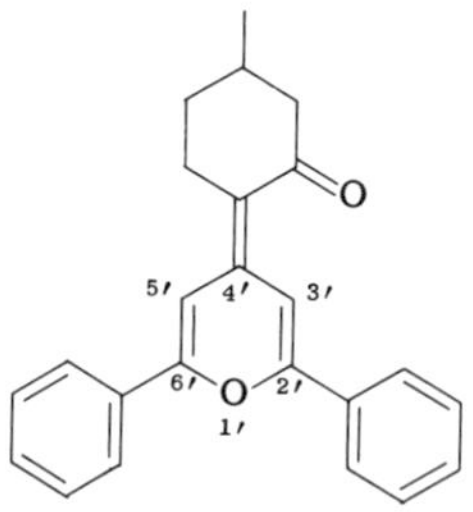

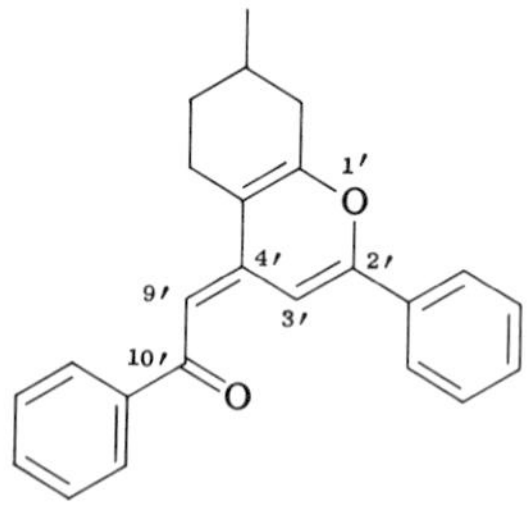

A synthesis of the isomers **8** and **9** was then undertaken. The sequence involves the reaction of a ketene (**11**), generated from the β-keto ester (**10**) with triphenylphosphinebenzoylmethylene (**12**) (Crabbé *et al.*, 1970). In addition to triphenylphosphine oxide, three reaction products are separated by chromatography. The first white crystalline substance shows spectral properties corresponding to the 7-methyltetrahydroflavone (**13**), with a molecular ion at m/e 240 (M^+), a UV spectrum exhibiting two absorption bands at λ_{max} 229 and 275 nm (log ϵ 4.15 and 4.40) and a

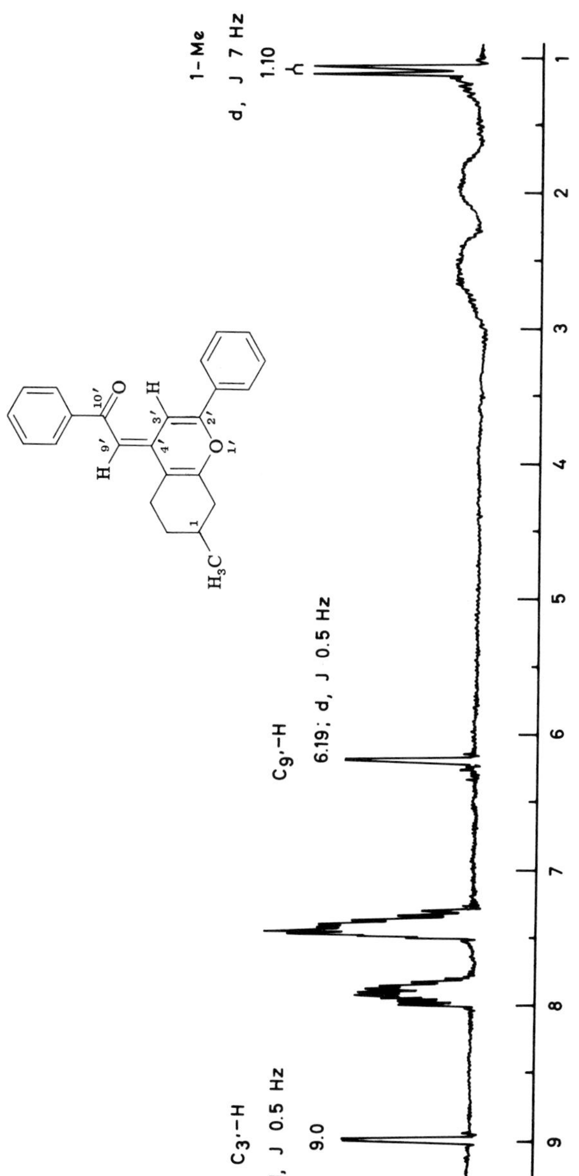

FIG. 1. Nuclear magnetic resonance spectrum of the substituted phenylmethylenepyran (**9**) in deuterochloroform solution.

typical vinylic proton appearing at 6.7 ppm. The second compound is a yellow crystalline substance, the physical properties of which agree with the 2′,6′-diphenylpyran structure (**8**). It analyzes for $C_{24}H_{22}O_2$ and shows the molecular ion m/e 342 (M^+). The intense UV bands (λ_{max} 221, 244, 307, 408 nm; log ϵ 4.30, 4.17, 4.19, 4.38) supports the conjugated system characterizing this molecule. In addition to ten aromatic H, the NMR of **8** shows an olefinic proton at C–5′ (6.65 ppm) which is coupled (J 2 Hz) with that at C–3′ (8.9 ppm), the latter being deshielded by the cyclohexanone carbonyl. The third reaction compound is the phenylmethylenepyran (**9**), identical in all respects, but for the optical properties, with the yellow substance isolated from the reaction of pulegone (**1**) with benzoyl chloride (see above). The strong deshielding of the vinylic proton at C–3′ in **9** is attributed to the C–10′ carbonyl. This chemical shift establishes the configuration of the double bond between C–4′ and C–9′ in the phenylmethylenepyran (**9**).

ϕ—C(=O)—CH=Pϕ_3

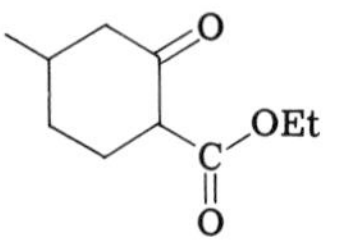

10 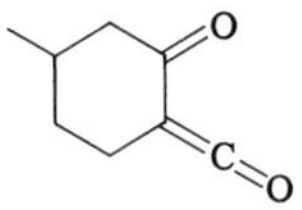 11 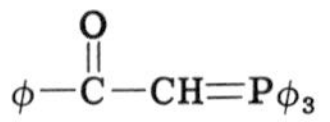 12 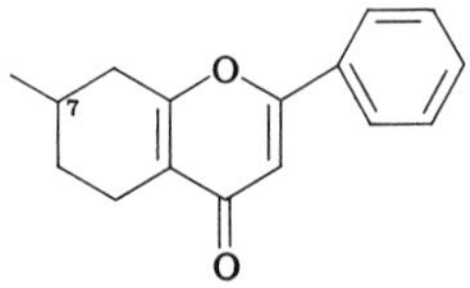 13

As illustrated in Fig. 2, the MS of isomers **8** and **9** show different fragmentation patterns under electron impact, which may be of some significance in structural studies.

The phenylmethylenepyran (**9**) is presumably formed from *d*-pulegone (**1**) through a dianion generated at C–9 and C–10, which reacts with benzoyl chloride, giving rise to a 1,5-diketone, which then cyclizes through the enol form of the cyclohexanone carbonyl (Crabbé *et al.*, 1970, 1972a).

The reaction of the phenylmethylenepyran (**9**) with hydroxylamine in ethanol-pyridine solution is of considerable interest, since it leads, in high yield, to the N-oxide (**14**) (Crabbé *et al.*, 1972b). The heterocyclic derivative **14** presents interesting spectroscopic properties. The MS shows a strong molecular ion m/e 372 (M^+), in agreement with the elementary analysis supporting the $C_{24}H_{24}O_2N_2$ formula. The UV spectrum displays one intense absorption maximum at λ_{max} 246 nm (log ϵ 4.53). The IR presents a hydroxyl absorption at 3400 cm^{-1}, and various intense bands at 1610, 1475, 1225, 1150, and 965 cm^{-1}, which appear to be typical of this heterocyclic system. Besides the doublet at 1.15 ppm due to the secondary methyl, and ten aromatic protons, the NMR shows a signal at 4.08 ppm

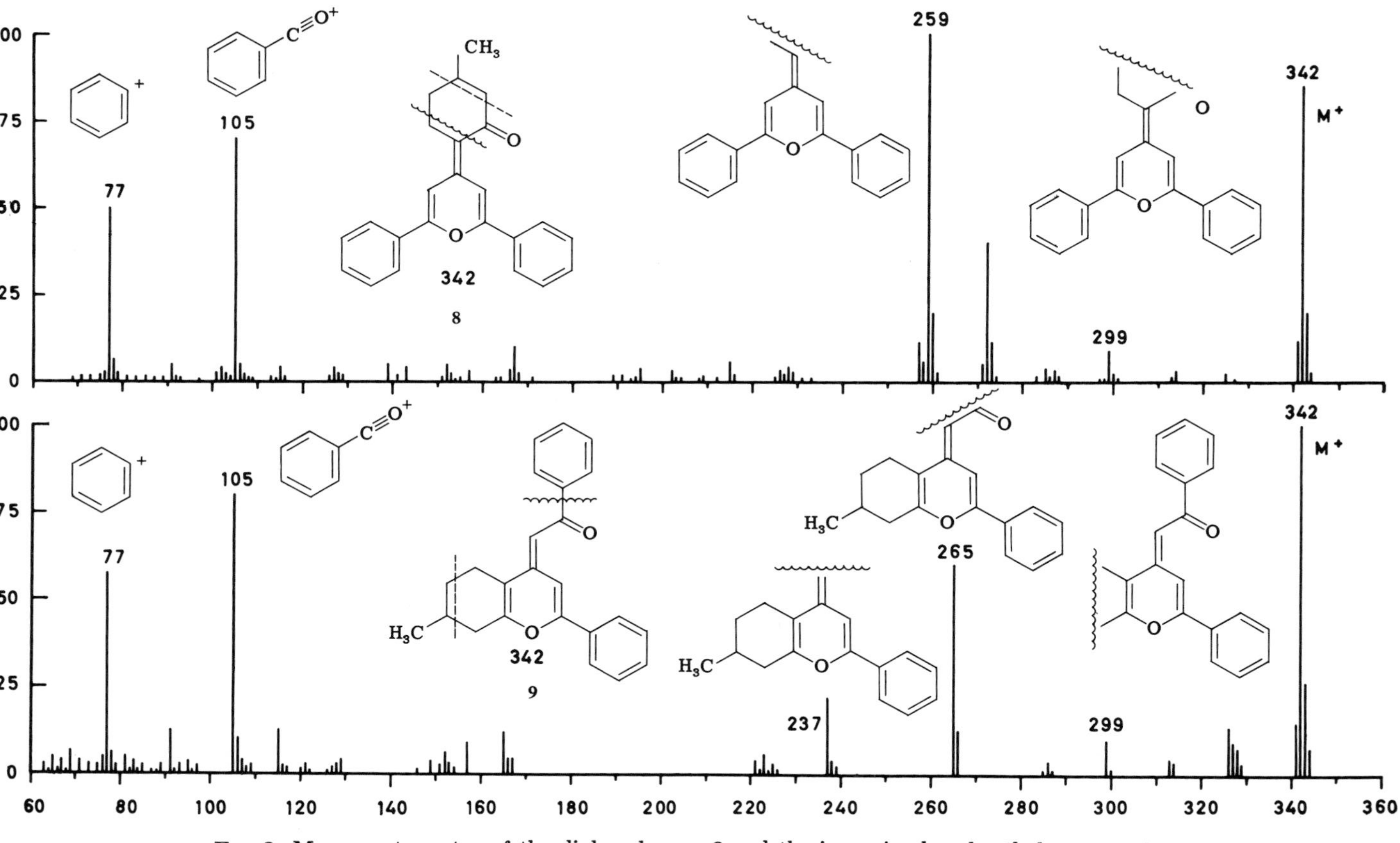

FIG. 2. Mass spectrometry of the diphenylpyran **8** and the isomeric phenylmethylenepyran **9**.

corresponding to the CH_2 group adjacent to the oxime, a 1-proton chemical shift at 6.94 ppm attributed to the proton in the heterocycle, and the oxime-hydroxyl signal at 10.05 ppm, readily exchanged with deutero-water.

It is significant to note that the course of the reaction of the methylenepyran (**9**) with hydroxylamine is different from that of the steroidal pyrone (**15**) under the same conditions (Crabbé *et al.*, 1971b). Indeed, one has shown that treatment of chromone **15** with hydroxylamine hydrochloride in pyridine solution leads to a mixture of isomeric steroids **16** and **17**. The 5′-arylisoxazole (**16**) is characterized by the isoxazole protons appearing as doublets (J 2 Hz) at 8.66 (C–3′H) and 7.30 ppm (C–4′H), respectively. The 3′-arylisoxazole isomer (**17**) shows signals at 7.40 (C–4′H) and 8.83 ppm (C–5′H), also as doublets (J 2 Hz).

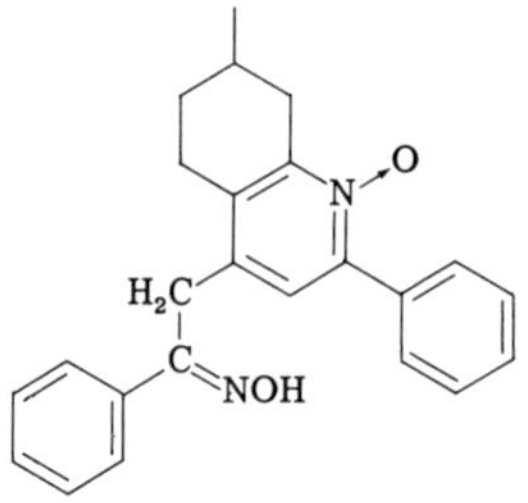

H_2C C=NOH N→O

14

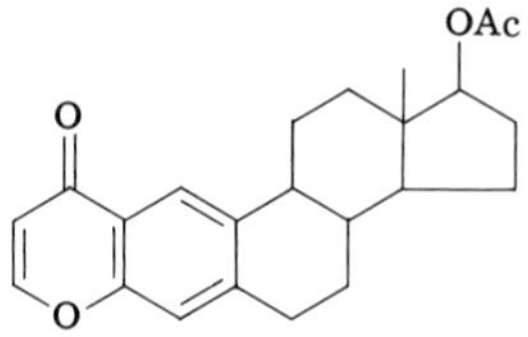

OAc O O

15

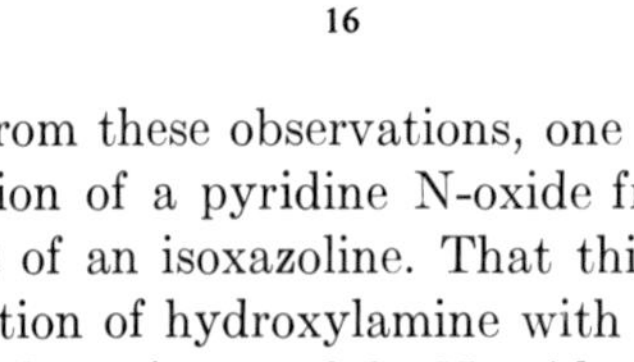

OAc 3′ 4′ 2′ N O 5′ 1′ HO

16

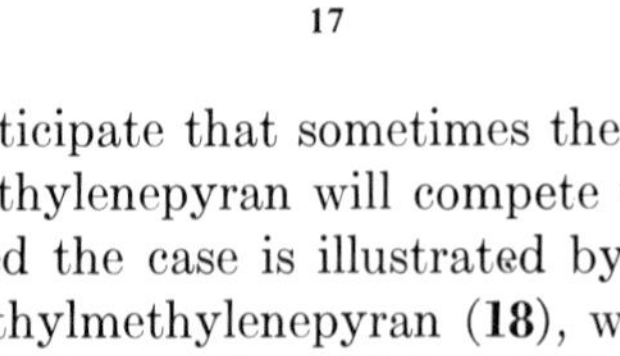

OAc 5′ 4′ 1′O 3′ N 2′ HO

17

From these observations, one would anticipate that sometimes the formation of a pyridine N-oxide from a methylenepyran will compete with that of an isoxazoline. That this is indeed the case is illustrated by the reaction of hydroxylamine with the dimethylmethylenepyran (**18**), which affords a mixture of the N-oxide (**19**) and the isoxazoline (**20**) in a 6:1 ratio. In the substituted pyridine oxide (**19**), both methyls appear at 2.48 ppm, whereas the aromatic protons give a single signal at 6.98 ppm. In addition, a typical UV spectrum (λ_{max} 216, 262 nm; log ϵ 4.47, 4.21) is associated with the pyridine N-oxide (**19**). In contrast, the isoxazoline (**20**), which has no UV absorption above 220 nm, does not present any olefinic proton, but three vinylic methyls at 1.72 and 2.18 ppm.

The determination of the complete structure and stereochemistry of these new compounds derived from pulegone constitutes an illustration of the power of the physical tools presently available in organic chemistry. Moreover, the foregoing discussion of the physical properties of the substances obtained by reaction of hydroxylamine with different systems has the feature of providing reasonable support for the proposed structures, in agreement with the reaction mechanisms involved. Hydroxylamine appears to be a versatile reagent which can lead to various interesting heterocyclic systems.

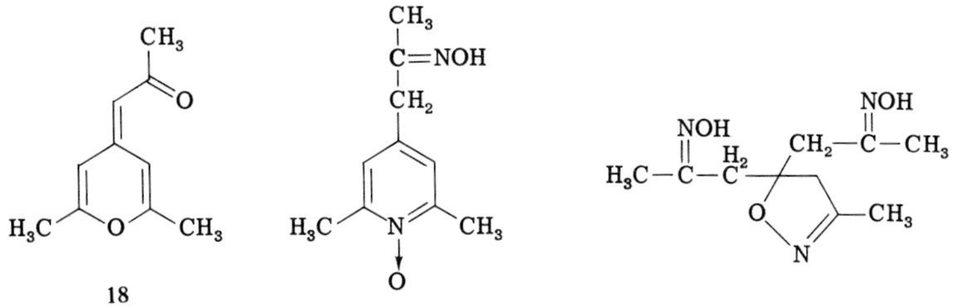

18 **19** **20**

Applications of Physical Methods to Steroid Problems

In this part are reported various applications of physical methods for the resolution of structural and configurational problems resulting from some new reactions which have been studied with steroids in the Syntex Research Laboratories in Mexico City.

The following example mentions an application of X-ray crystallography for the assignment of configuration to a pentacyclic steroid. Photochemical addition of *cis*- and *trans*-dichloroethylene to the Δ^{16}-20-ketosteroid (**21**) affords two α-face adducts which differ only in the stereochemistry of the chlorine atoms (Sunder-Plassman *et al.*, 1969).

The configuration of the 17′-chlorines is considered to be *endo* in compound **22a** and *exo* in the other isomer **23** on the basis of the observed long-range coupling (J 1.5 Hz) between the 16β and 17′β proton in compound **22a**, which are in a "W" or "M" spatial relationship to each other. This long-range coupling is absent in compound **23** (Sunder-Plassman *et al.*, 1969). Since the configuration at the 16′ position could not be assigned with confidence on the basis of the $J_{16\beta,16'}$ and $J_{16',17'}$ values alone, the 3β-bromoacetate derivative (**22b**) was subjected to X-ray analysis. The X-ray data of the 3β-bromoacetoxydichloro steroid ($C_{25}H_{33}O_3Cl_2Br$; orthorhombic crystals with a = 32.37, b = 9.77, and c = 7.89 Å; space group $p2_12_12_1$ from systematic absences) support the presence of a 16′-*exo*-

chlorine in compound **22b** (Christensen *et al.*, 1971). The cyclobutane ring in **22b** is planar to within 0.015 Å with 110° and 119° dihedral angles between the 16β,16′α and 16′α,17′β protons.

The stereochemistry of the 16′-chlorine in compound **23** can be deduced by comparison of the $J_{16\beta H,16'H}$ (9.5 *viz.* 4.5 Hz) and $J_{16'H,17'H}$ values (7.5 *viz.* 6.0 Hz) in both isomers. These values indicate a difference in the relative configuration of the 16β,16′-protons which is *trans* in steroids (**22**) and therefore has to be *cis* with a very small dihedral angle in compound **23**. Consequently, the configuration of the 16′-chlorine is *endo* in compound **23**, which is consistent with the observed *trans* relationship of the 16′,17′-protons in both isomers.

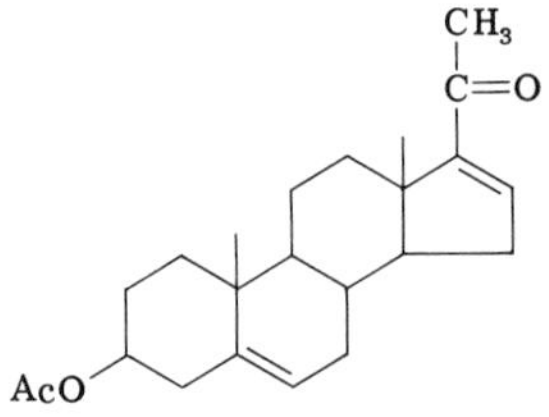

CH3 C=O AcO

21

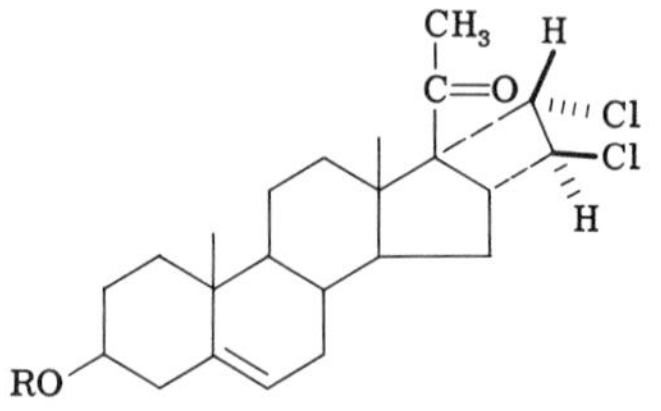

CH3 H C=O Cl Cl H RO

22a. R = Ac
b. R = BrCH$_2$—C(=O)—

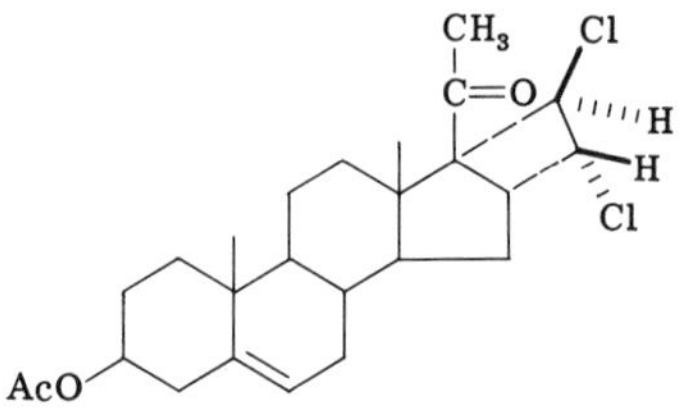

CH3 Cl C=O H H Cl AcO

23

The fact that both *cis*- and *trans*-dichloroethylene provide the same product composition is in agreement with the stepwise addition mechanism forming a diradical intermediate which can undergo free rotation at the 16′,17′-bond before ring closure.

The chemistry of difluorocarbenes leads to a wide variety of useful reactions. Some new interesting applications in organic synthesis will be reviewed briefly.

Addition of difluorocarbene to the enol acetate (**24**) yields the 2α,3α-difluorocyclopropyl steroid (**25**) (Crabbé *et al.*, 1972c). This substance is typified by an IR which shows the 3β-acetoxy absorption at 1758 cm^{-1} and the 17-acetate at 1728 cm^{-1}. In addition, the α-configuration of the difluorocyclopropane ring is evidenced by the absence of long-range coupling between fluorine and 19-methyl protons. Treatment of compound **25** with base gives the saturated A-homo-difluoro-ketosteroid (**26**) in good yield, with no evidence of the presence of a conjugated ketone. This indicates the course of the homologation reaction to be substantially different from that of dichloro and dibromo cyclopropanes, known to afford exclusively conjugated enones. Although cycloheptanones usually show an IR band in the 1700 cm^{-1} region, the difluoro ketone (**26**) is characterized by an intense carbonyl absorption at 1744 cm^{-1}. This shift of the IR absorption band is attributed to the α-fluorine atoms adjacent to the carbonyl group. In contrast to the above result, addition of difluorocarbene to the cyclopentanone enol acetate (**27**) allows one to isolate only a small amount of the pentacyclic steroid (**28**), the major product being the *D*-homo-steroid (**29a**) resulting from ring opening and elimination of HF. Base treatment of compound **28** gives exclusively the fluoro-enone (**29b**) characterized by the

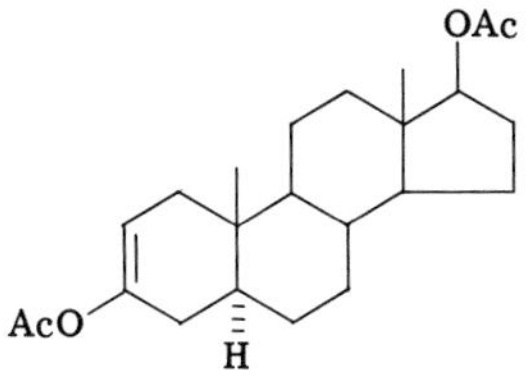

24

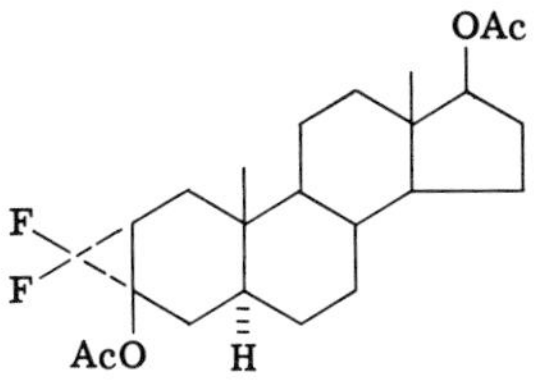

25

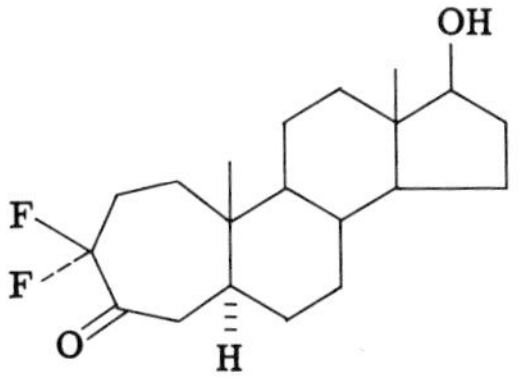

26

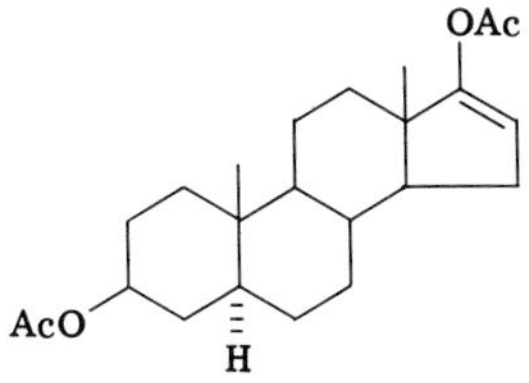

27

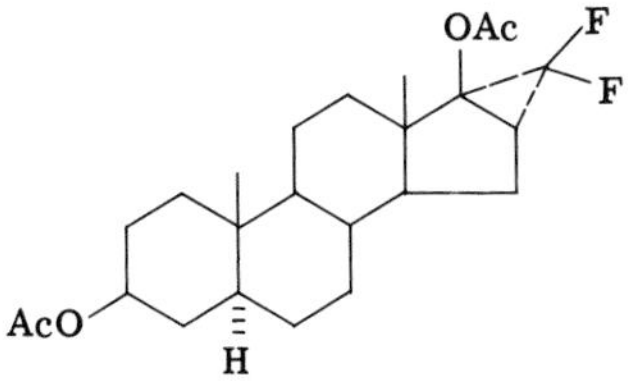

28

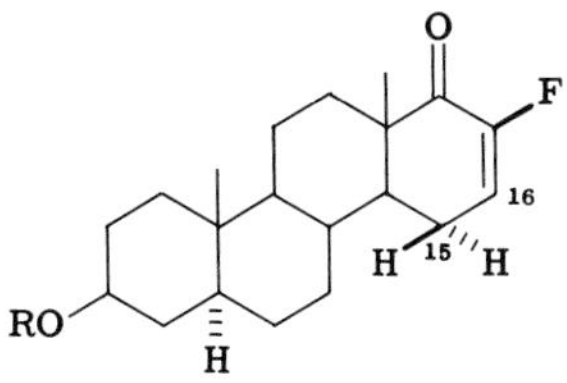

29a. R = Ac
b. R = H

UV absorption at 234 nm (log ϵ 3.85) and the IR bands at 1690 and 1665 cm^{-1}, supporting the presence of an α,β-unsaturated carbonyl. Finally, the C–16 vinylic proton appears as a multiplet centered at 6.18 ppm ($J_{15\alpha H,16H}$ 6 Hz; $J_{15\beta H,16H}$ 3 Hz; $J_{16H,16\alpha F}$ 15 Hz). These results seem to indicate that the course of the ring expansion reaction is highly dependent on the conformational strain of the cyclic system.

Difluorocarbene addition to the 17α-acetylenic group of **30** provides the difluorocyclopropene derivative (**31a**) (Velarde *et al.*, 1970). Alkaline hydrolysis followed by mild reacetylation gives the monoacetate (**31b**). Reaction of the difluorocyclopropenyl carbinol (**31b**) with 2-chloro-1,1,2-trifluorotriethylamine affords a mixture of three isomeric substances, as evidenced by their elementary analysis and identical molecular ions (*m/e* 410 M^+) (Crabbé *et al.*, 1971a). From previous experience with the fluoramine reagent, the first compound is the 17β-fluorosteroid (**31c**), which is typified by the vinylic proton at 7.5 ppm, appearing as a triplet, because it is coupled with two nonequivalent fluorines. The second substance is the trifluoromethyl allene (**32a**), characterized by the propadiene band at 1980 cm^{-1} in the IR. The third compound is its C–21 isomer (**32b**) also typified by an allene band at 1980 cm^{-1}, but differing from **32a** by the position of the 18-angular methyl, which appears at 0.925 ppm in **32a** and 0.861 ppm in **32b**.

Reaction of 3β,17β-dihydroxy 17α-difluorocyclopropenyl 5α-androstan-3-acetate (**31b**) with formic acid affords the 17α-cyclopropenonyl steroid

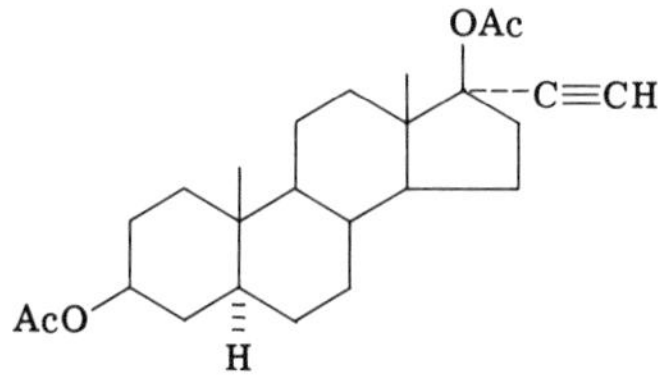

30

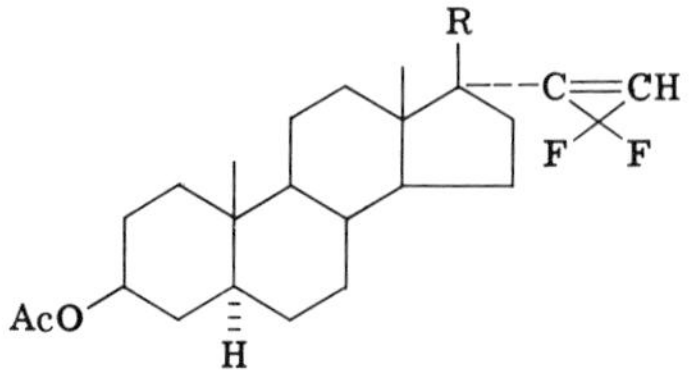

31a. R = OAc
b. R = OH
c. R = F

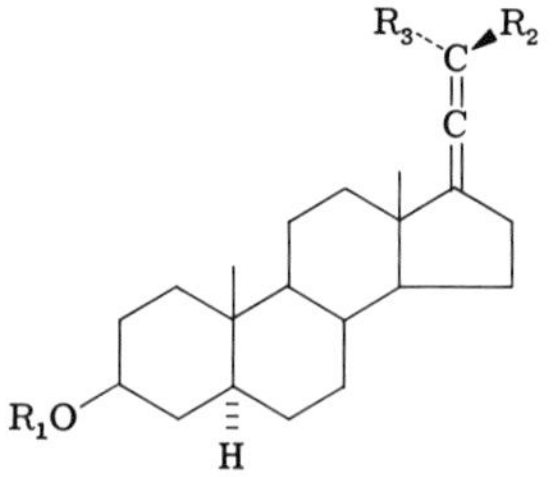

32a. R_1 = Ac, R_2 = CF_3, R_3 = H
b. R_1 = Ac, R_2 = H, R_3 = CF_3
c. R_1 = Ac, R_2 = COF, R_3 = H
d. R_1 = H, R_2 = CO_2Me, R_3 = H
e. R_1 = Ac, R_2 = $COCH_3$, R_3 = H
f. R_1 = Ac, R_2 = H, R_3 = $COCH_3$
g. R_1 = Ac or H, R_2 = R_3 = H

(**33a**). Mild alkaline hydrolysis of **33a** gives the corresponding 3,17-diol (**33b**). The physical properties of **33b** and its position isomer **34** are in agreement with the aromatic character of the cyclopropenone chromophore. The signal of the cyclopropenonyl vinylic proton appears at 8.40 ppm in the NMR. Compounds **33b** and **34** show IR bands in the region of 1840, 1635, and 1580 cm^{-1}. Both steroids also present a low-intensity UV absorption at ca. 260 nm, i.e., the cyclopropenone (**33b**), at λ_{max} 259 nm ($\log \epsilon$ 1.77) and its isomer (**34**), at λ_{max} 262 nm ($\log \epsilon$ 1.64), in addition to a more intense band at 225 nm ($\log \epsilon$ 2.15).

When the cyclopropenonyl chromophore is located in a dissymmetric surrounding, it displays Cotton effects. Figure 3, which reproduces the UV and CD curves of the three steroidal cyclopropenones **33b**, **34**, and **35**,

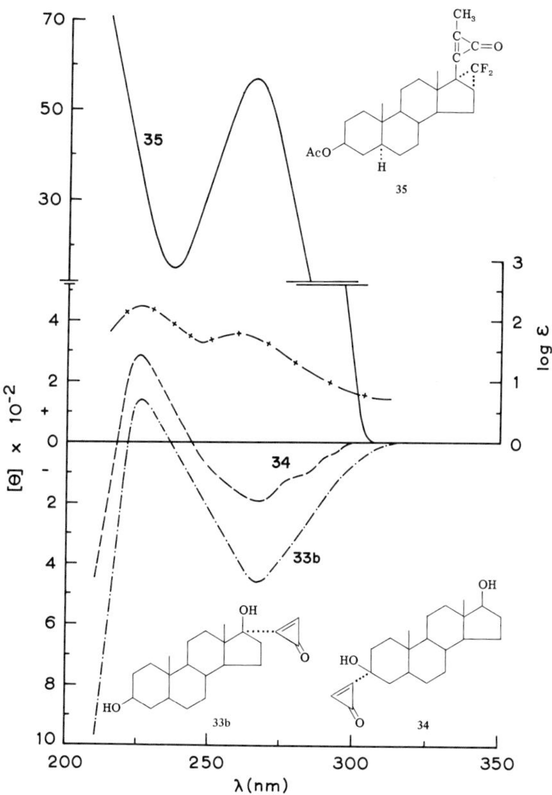

FIG. 3. Ultraviolet and circular dichroism curves of the isomeric steroidal cyclopropenones (**33**b) and (**34**) and of the 17β-cyclopropenonyl steroid (**35**).

clearly shows both transitions to be optically active. Whereas the three steroids **33b**, **34**, and **35** exhibit a positive Cotton effect at ca. 220 nm, a weak negative Cotton effect is associated with the cyclopropenone chromophore around 265 nm in steroid **33b** and in its 3α-isomer (**34**). Conversely, the 17β-cyclopropenonyl steroid (**35**), displays an intense positive CD curve in the same spectral region. The molecular ellipticity of the 17α-cyclopropenonyl steroid (**33b**) ($[\theta]_{266}$ − 460), is slightly more intense than that of its 3α-analog, **34** ($[\theta]_{268}$ − 190). This can be attributed to steric hindrance at C–17, leaving less conformational freedom to the cyclopropenone group than at position 3. For similar reasons, the intensity of the Cotton effect associated with the 17β-cyclopropenonyl steroid (**35**) ($[\theta]_{265}$ + 5780), is much higher than in either **33b** or **34**. In compound **35**, the 18-angular methyl and the 16,17α-difluoromethylene function freeze the conformation of the 17β-chain, thus enhancing substantially the Cotton effect. Finally, the sign of the Cotton effect associated with the 265 nm transition in the 17α- (**33b**) and 17β-cyclopropenonyl (**35**) steroids is the same as that of 17-acetyl steroids presenting an identical configuration.

Treatment of the carbinol **33a** with 2-chloro-1,1,2-trifluorotriethylamine gives the allenic acid fluoride **32c**. It is characterized by a UV presenting an absorption at λ_{max} 226 nm (log ε 4.23). Its IR shows the allene band at 1960 cm^{-1} and the acid fluoride function at 1810 cm^{-1}, besides the acetate at 1730 cm^{-1}. The NMR exhibits a triplet at 5.51 ppm (J 4 Hz) corresponding to the 21-allenyl proton. When **32c** is allowed to react with sodium

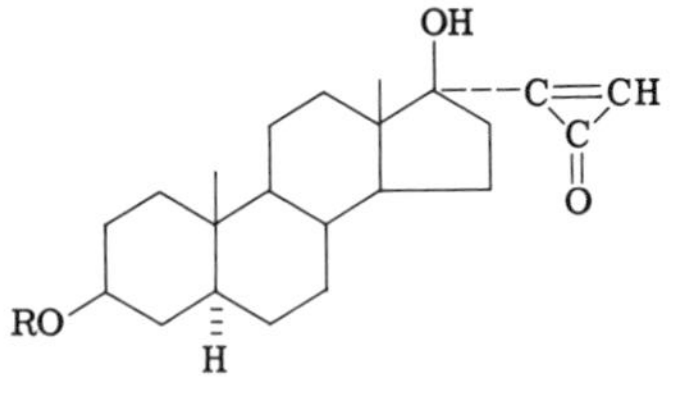

33a. R = Ac
b. R = H

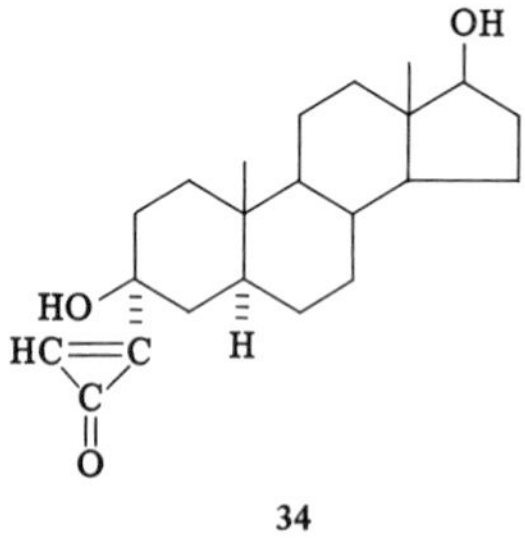

34

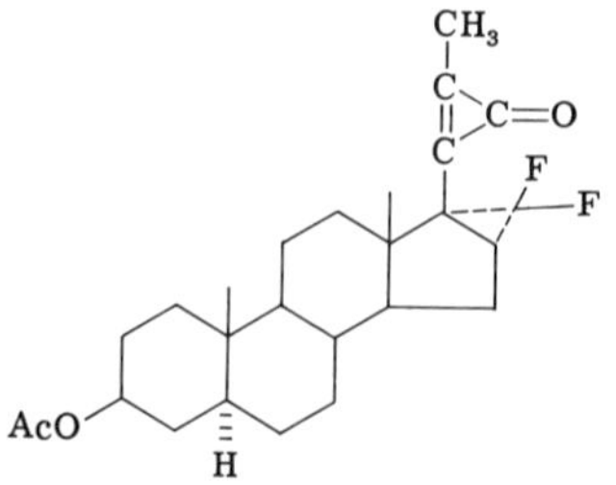

35

methoxide, it is converted into its methyl ester (**32d**). Further treatment of **32d** with sodium methoxide gives the enol ether (**36**) by a Michael type addition on the central carbon atom of the allenyl group. Acid hydrolysis of **36** yields the ketone **37a**, thus making the above sequence a novel synthesis of β-ketoesters. Treatment of **37a** with methanolic potassium hydroxide at reflux cleaves the β-keto-ester group and affords the pregnane derivative **37b** in high yield (Crabbé *et al.*, 1971a).

Similarly, reaction of lithium dimethyl copper on the allenic acid fluoride (**32c**) provides the ketone (**38**), characterized by the absence of allene band in the IR and the appearance of a new vinylic methyl signal at 1.6 ppm in the NMR, besides the methyl ketone chemical shift at 2.13 ppm (Crabbé and Velarde, 1971). As in the case of compound **36**, the $LiMe_2Cu$ reagent also attacked the central carbon atom of the allene system. In contrast, treatment of the acid fluoride, **32c**, with dimethyl cadmium gives a mixture of isomeric allenic methyl ketones **32e** and **32f**. During this reaction the propadiene moiety is not affected. This is confirmed by the presence of an allene band at 1940 cm^{-1} in the IR, which also shows the conjugated keto band at 1670 cm^{-1}. The allenic methyl ketones, **32e** and **32f**, exhibit a typical UV absorption at 232 nm. Finally, the NMR clearly shows the methyl ketone signal at ca. 2.15 ppm, in addition to the 21-allenyl proton chemical shift around 5.7 ppm (Crabbé and Velarde, unpublished observations).

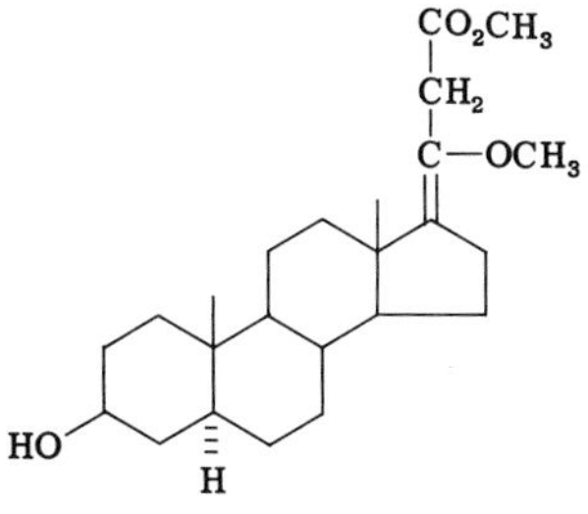

36

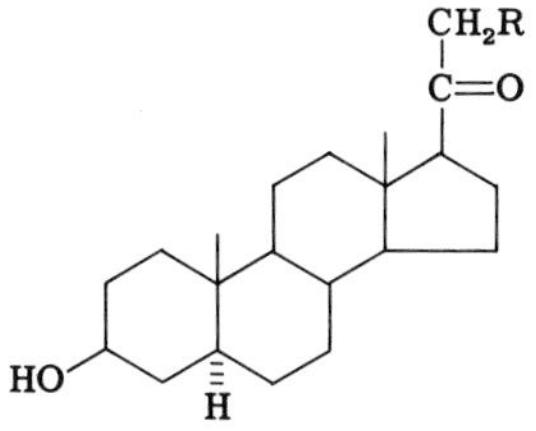

37a. R = CO_2CH_3
b. R = H

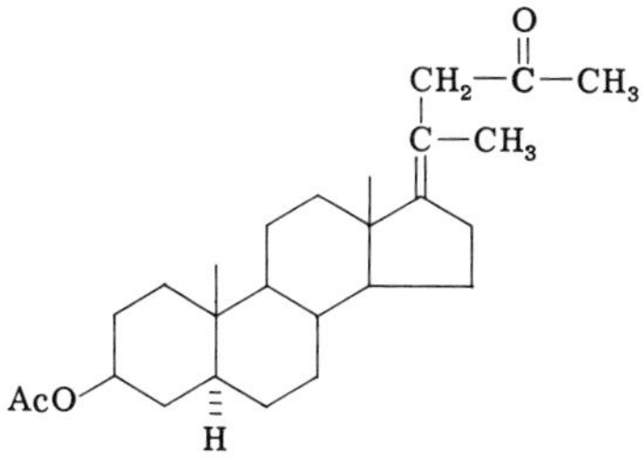

38

An important step in the total synthesis of steroids is the elaboration of the pregnane and corticoid chains at C–17. A two-step synthesis of these chains is easily achieved by taking advantage of allene chemistry.

The allenyl steroid (**32g**) is prepared in almost quantitative yield by reaction of the 17α-ethynyl 17β-acetate (**30**) with zinc dust in refluxing diglyme (Biollaz *et al.*, 1972), by a reduction process, accompanied by rearrangement and elimination of the 17β-acetoxy group. The propadienyl steroid (**32g**) is characterized by IR absorption at 1960 cm^{-1} and the 21-protons NMR signal at 4.66 ppm. Whereas, exposure of **32g** to *m*-chloroperbenzoic acid provides a mixture of the pregnan-20-one (**39**) and the corticoid ester (**40a**), reaction of **32g** with osmium tetroxide-pyridine, followed by cleavage of the osmate ester with sodium sulfite and potassium bicarbonate affords exclusively the corticoid **40b**. The structure and stereochemistry of both **39** and **40** are easily defined by conversion into known steroids (Biollaz *et al.*, 1972).

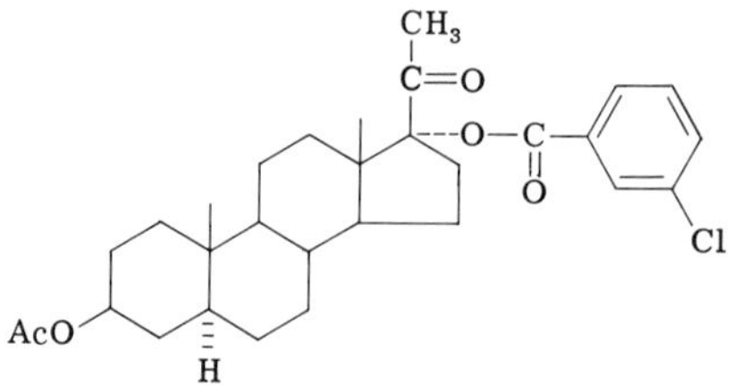

39

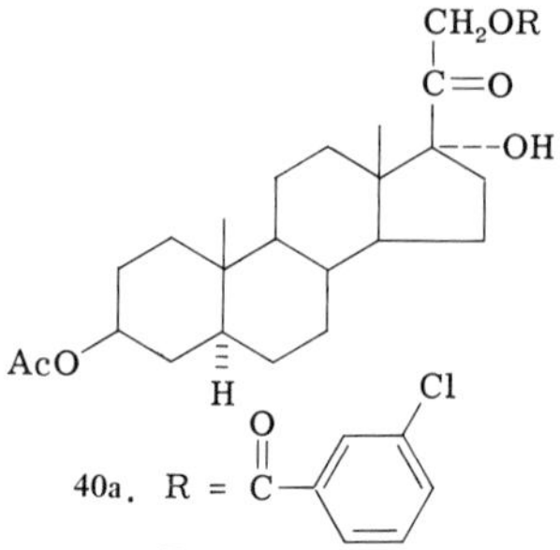

40a. R = $C(=O)C_6H_4Cl$

b. R = H

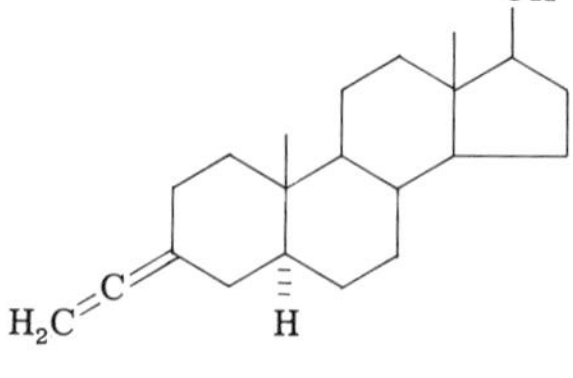

41

It has been established that mono- and bisepoxides are intermediates in such reactions. Hence, the pregnan-20-one (**39**) and the corticoid (**40**) chains are the opening products of a 17,20-allene oxide intermediate and a dioxaspiro [2.2]pentane precursor, respectively (Biollaz *et al.*, 1972).

Since only few methods are available for the assignment of stereochemistry of the allene chromophore, we decided to investigate the chirality of the propadiene system by CD. A joint research effort with Professor

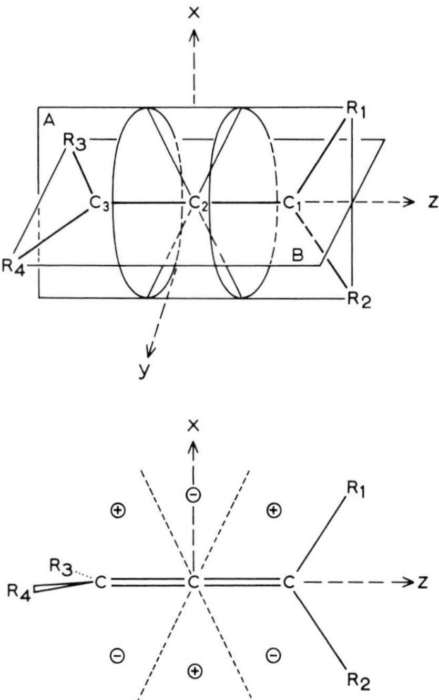

Fig. 4. The bifurcated quadrant rule for optically active allenes. The most polarizable substituent (R_1) is aligned uppermost along a vertical axis. If the more polarizable of the two rear substituents in the horizontal plane is R_3, the lowest energy (long wavelength) Cotton effect is positive. Conversely, if it has the R_4 configuration, the Cotton effect is negative (Crabbé *et al.*, 1972d).

Mason from the King's College and Professor Moore of the Massachusetts Institute of Technology, has led to proposing a bifurcated quadrant rule for the allene group (Crabbé *et al.*, 1971c). This rule establishes a correlation between the sign of the Cotton effect and the chirality of the propadiene chromophore. The signs indicated in Fig. 4 refer to the lowest energy Cotton effect associatee with groups substituted into the +Y hemisphere. The signs are opposite for the groups substituted into the −Y hemisphere.

The bifurcated quadrant rule, shown in Fig. 4, is in agreement with the CD data obtained for the 230–250 nm transition of a variety of mono-, di-, tri-, and tetra-substituted steroidal allenes prepared recently in our laboratory.

As shown in Fig. 5, the disubstituted 3-allenyl steroid (**41**) presents a positive Cotton effect ($[\theta]_{229}$ + 3760). Ring B falls in the upper right quadrant, leading to a positive Cotton effect, in agreement with the rule.

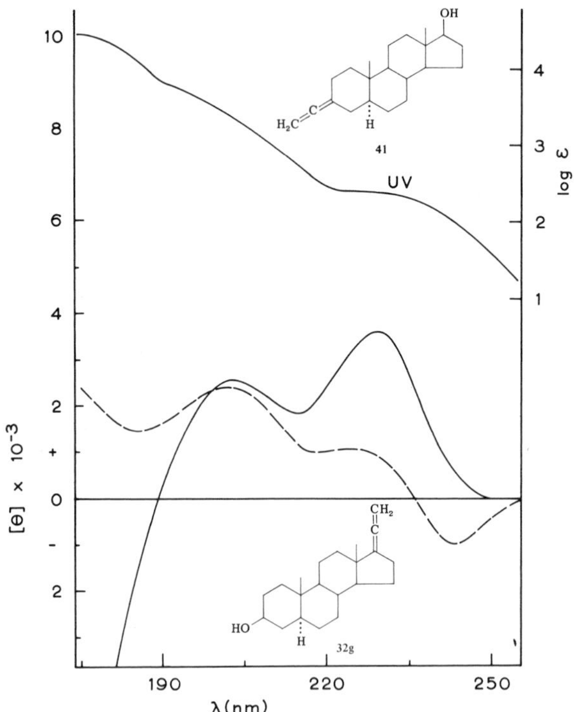

FIG. 5. Ultraviolet and circular dichroism curves of the 3-allenyl steroid (**41**) and its 17-isomer (**32g**).

In the same spectral region the 17-allenyl steroid (**32g**) exhibits a weakly negative Cotton effect ($[\theta]_{243}$ − 920). The bifurcated quadrant rule predicts that the negative contribution of the angular methyl at C–13 will be partially cancelled by the positive influence of ring C, in agreement with experience.

Figures 3 and 5 clearly indicate that in addition to revealing useful stereochemical features, sometimes the chiroptical methods will also provide interesting spectroscopic information, i.e., several transitions hardly detected by UV technique are clearly apparent on these CD curves.

ACKNOWLEDGMENTS

I should like to thank my co-workers from the UNAM, especially Mr. J. Haro, Dr. L. A. Maldonado, Mr. C. Ríus, and Dr. E. Santos for their excellent work and cooperation. I also want to express my sincere gratitude to my colleagues at Syntex, in particular Misses A. Cruz and E. Velarde, as well as Dr. J. Iriarte and two postdoctors, Drs. M. Biollaz and W. Haefliger, for their outstanding contribution and careful observations.

Bibliography

Some general references on applications of physical methods in organic chemistry:

Bhacca, N. S., and D. H. Williams. 1964. "Applications of NMR Spectroscopy in Organic Chemistry." Holden-Day, San Francisco, California.

Brand, J. C. D., and G. Eglinton. 1965. "Applications of Spectroscopy to Organic Chemistry." Oldbourne Press, London.

Budzickiewicz, H., C. Djerassi, and D. H. Williams. 1967. "Mass Spectrometry of Organic Compounds." Holden-Day, San Francisco, California.

Crabbé, P. 1968. "Applications de la Dispersion Rotatoire Optique et du Dichroïsme Circulaire en Chimie Organique." Gauthier-Villars, Paris.

Crabbé, P. 1972. "Chiroptical Methods in Chemistry: An Introduction." Academic Press, New York.

Dyer, J. R. 1965. "Applications of Absorption Spectroscopy of Organic Compounds." Prentice-Hall, Englewood Cliffs, New Jersey.

Fay, M., G. R. Mallet, and W. M. Mueller. 1966. "Advances in X-Ray Analysis." Plenum, New York.

Jackman, L. M., and S. Sternhill. 1969. "Applications of Nuclear Magnetic Resonance Spectroscopy in Organic Chemistry." Pergamon, Oxford.

Schwarz, J. C. P. 1964. "Physical Methods in Organic Chemistry." Holden-Day, San Francisco, California.

References

Biollaz, M., W. Haefliger, E. Velarde, P. Crabbé, and J. H. Fried. 1972. *Chem. Commun.*

Christensen, A., P. Crabbé, A. Cruz, and L. Tökes. 1971. *J. Org. Chem.* **36**:2381.

Crabbé, P. 1954. Ph.D. Thesis, Univ. of Paris.

Crabbé, P., and E. Velarde. 1972. *Chem. Commun.* p. 241.

Crabbé, P., E. Díaz, J. Haro, G. Pérez, D. Salgado, and E. Santos. 1970. *Tetrahedron Lett.* p. 5069.

Crabbé, P., H. Carpio, and E. Velarde. 1971a. *Chem. Commun.* p. 1028.

Crabbé, P., L. A. Maldonado, and I. Sánchez. 1971b. *Tetrahedron* **27**:711.

Crabbé, P., E. Velarde, H. W. Anderson, S. D. Clark, W. R. Moore, A. F. Drake, and S. F. Mason. 1971c. *Chem. Commun.* p. 1261.

Crabbé, P., E. Díaz, J. Haro, G. Pérez, D. Salgado, and E. Santos. 1972a. *J. Chem. Soc.* p. 41.

Crabbé, P., J. Haro, C. Ríus, and E. Santos. 1972b. Submitted for publication.

Crabbé, P., A. Cervantes, L. Cuéllar, A. Cruz, E. Galeazzi, and J. Iriarte. 1972c. Submitted for publication.

Nakanishi, K. 1964. "Infrared Absorption Spectroscopy." Holden-Day, San Francisco, California.

Sunder-Plassman, P., P. H. Nelson, P. H. Boyle, A. Cruz, J. Iriarte, P. Crabbé, J. A. Zderic, J. A. Edwards, and J. H. Fried. 1969. *J. Org. Chem.* **34**:3779.

Velarde, E., P. Crabbé, A. Christensen, L. Tökes, J. W. Murphy, and J. H. Fried. 1970. *Chem. Commun.* p. 725.

NOVEL SESQUITERPENES ISOLATED IN COMPOSITAE

J. ROMO, A. ROMO DE VIVAR, and A. ORTEGA

Instituto de Quimica de la Universidad Nacional Autónoma de México, México 20, D.F., México

Introduction

The germacranolide series of sesquiterpene lactones is continuously being enlarged by the addition of new members. Subgroups possessing certain common characteristics can now be formed from the large number of germacranolides presently known. A major subgroup of the germacranolides consists of those with the five-membered lactone group oriented to either C-6 or C-8. For example, pyrethrosin (**1**), one of the earliest known germacranolides (Barton and de Mayo, 1957), possesses a lactone oriented to C-8, whereas the lactone grouping of costunolide (**2**) is closed to C-6 (Rao *et al.*, 1960). The germacranolides are formed biogenetically by cyclization of farnesyl pyrophosphate. The resulting products contain two double bonds at C-1(10) and at C-4. These functions, which are present in a large number of germacranolides, permit their biogenetic transformation into eudesmanolides or guaianolides. Biogenetic oxidation of the double bonds at C-1(10) and C-4 produces epoxide functions as in pyrethrosin (**1**) and parthenolide (**3**) (Sousek *et al.*, 1961). Oxidation of the allylic positions to the C-1(10) and C-4 double bonds is frequently observed as in artemisiifolin (**4**). In isabelin (**5**) (Yoshioka *et al.*, 1968; Yoshioka and Mabry, 1969) and elephantopin (**6**) (Kupchan *et al.*, 1966), one of the methyl groups is oxidized to a carboxyl lactonized with a secondary hydroxyl group.

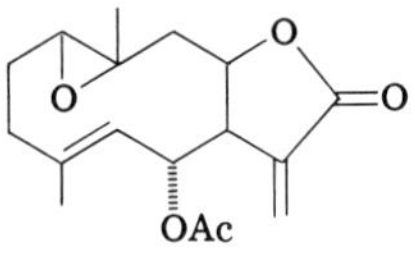

OAc

Pyrethrosin

1

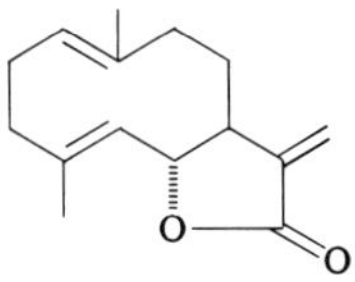

Costunolide

2

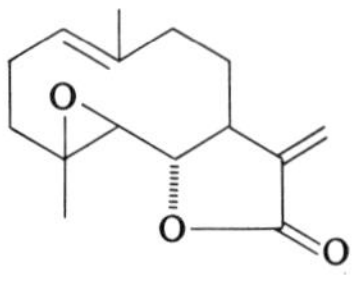

Parthenolide

3

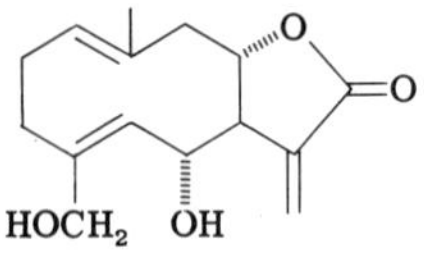

H_2C CH_3

HOCH$_2$ OH

Artemisiifolin

4

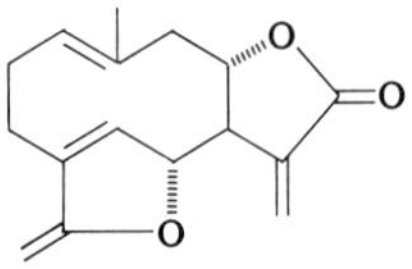

Isabelin

5

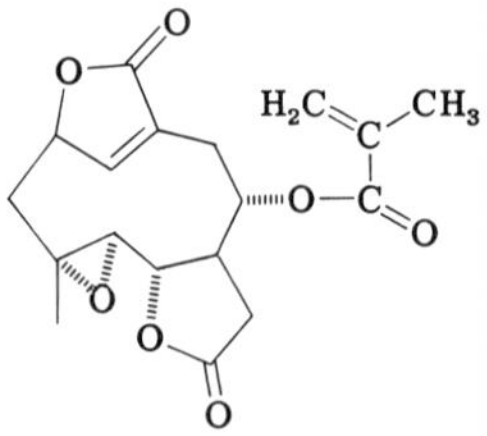

Elephantopin

6

Several germacranolides containing the C-1(10) and C-4 double bonds undergo the Cope rearrangement. This reaction is usually carried out by heating the germacranolides at relatively high temperatures.

Saussurea lactone (**7**) was initially considered a constituent of *Saussurea lappa* (Rao *et al.*, 1961); however, it was later demonstrated that it was produced from costunolide (**2**) in the process of purification of the extract of the plant. Compound **7** possesses the ethenyl and isopropenyl side chains previously found in the elemenes, sesquiterpene hydrocarbons isolated from various natural sources. Other germacranolides are transformed by heat into products containing the elemene system. There is some evidence that the Cope rearrangement also occurs under mild conditions (Benesová *et al.*, 1970).

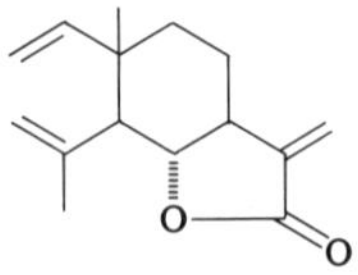

Saussurea lactone

7

Compounds Extracted from Compositae

We have recently examined the extract of *Zinnia acerosa*, a member of the Compositae widely distributed in the arid zones of the northern High Plateau of Mexico (Romo *et al.*, 1971). Zinarosin (**8**) and dihydrozinarosin (**9a**) were found as constituents of plants collected in May, 1969; (**9a**) was isolated as the diacetate (**9b**).

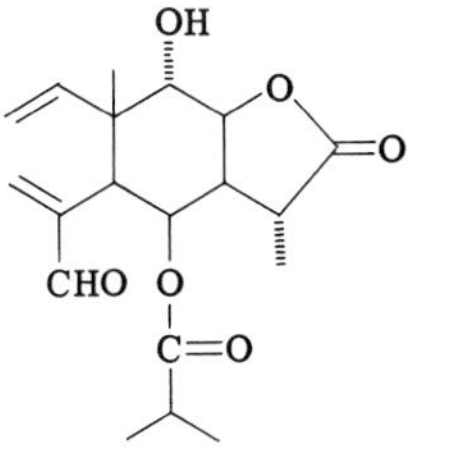

Zinarosin

8

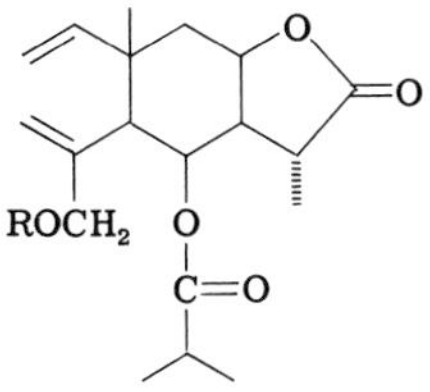

9a. R = H
b. R = Ac

These products were isolated using mild conditions which render unlikely a rearrangement of precursor germacranolides, such as **10a** or **10b**.

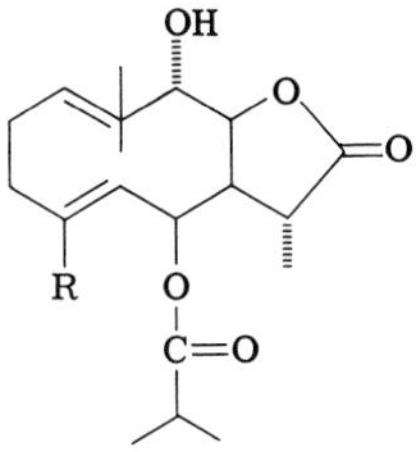

10a. R = —CH=O
b. R = —CH_2OH

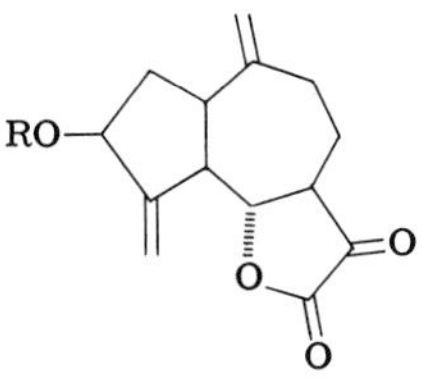

Zaluzanin

11a. R = H
b. R = Ac

From a plant collected in March, 1967, we isolated only zaluzanin C (**11a**) and its acetate, zaluzanin D (**11b**), guaianolides previously found in various species of *Zaluzania* (Romo de Vivar *et al.*, 1967). Further studies in the genus *Zinnia* are presently being carried out in our laboratories.

Examination of the extract of *Zexmenia brevifolia*, a member of the Compositae occurring in northern Mexico, led to the isolation of several sesquiterpenes which we named zexbrevin and zexbrevins B and C, all of which bear biosynthetic interrelationships, not only among themselves, but also with other germacranolides isolated from different species of Compositae. The structure of zexbrevin (**12**), the first member of this subgroup was established by Romo de Vivar *et al.* (1970). This germacranolide contains a 3(2*H*)-furanone ring, a C-6 oriented five-membered lactone, and a C-8 hydroxyl group esterified by methacrylic acid.

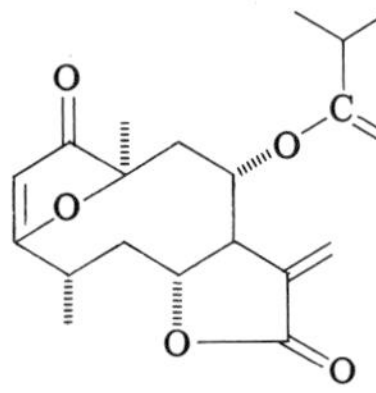

Zexbrevin

12

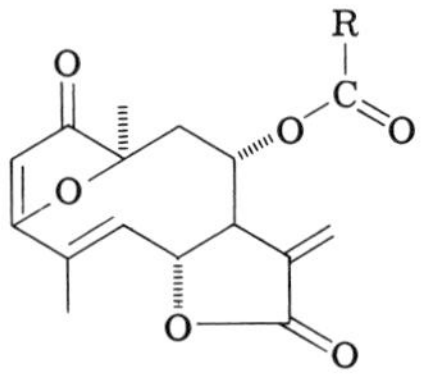

13a. R = $-C\!\begin{smallmatrix}=CH_2\\ -CH_3\end{smallmatrix}$ (Calaxin)

b. R = $-CH(CH_3)_2$ (Ciliarin)

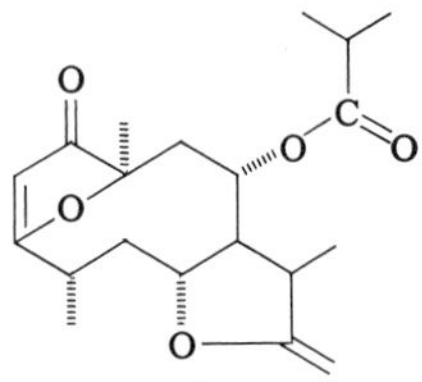

Tetrahydrozexbrevin

14

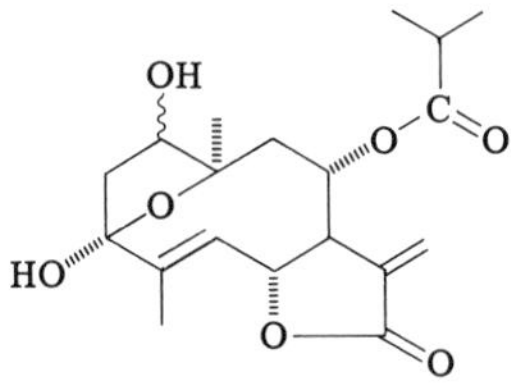

15a. R = $-C\!\begin{smallmatrix}=CH_2\\ -CH_3\end{smallmatrix}$ (Zexbrevin B)

b. R = $-CH(CH_3)_2$ (Orizabin)

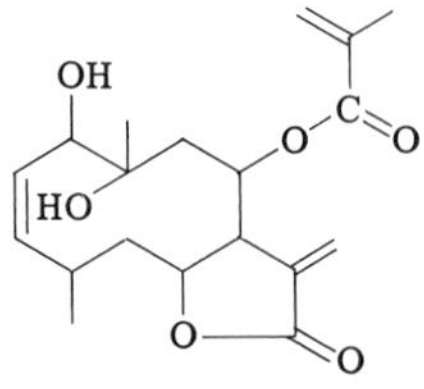

Zexbrevin C

16

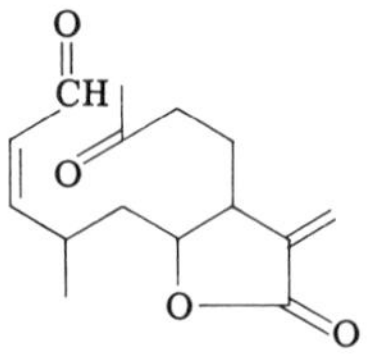

17

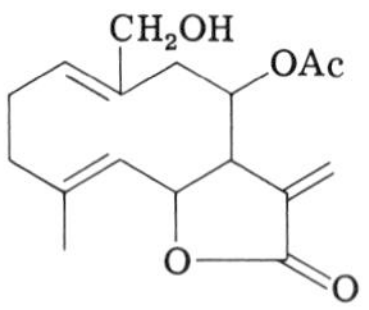

Zexbrevin D

18

A 3(2*H*)-furanone germacranolide, calaxin (**13a**), was isolated from *Calea axillaris* DC; and ciliarin (**13b**), from *Helianthus ciliaris* DC (Ortega *et al.*, 1970). Compounds **13a** and **13b** both contain a C-4 double bond and differ only in the nature of the acid which esterifies the C-8 hydroxyl group. In calaxin (**13a**) the esterifying moiety is methacrylic acid, and in ciliarin (**13b**) it is isobutyric acid. Correlation of these germacranolides with zexbrevin (**12**) was achieved when the hexahydro derivative of calaxin and tetrahydrociliarin were identified with tetrahydrozexbrevin (**14**).

The paramagnetic induced shifts observed in the nuclear magnetic resonance (NMR) spectra of zexbrevin (**12**) and calaxin (**13a**) in the presence of tris(dipivalomethanate) europium (III) [Eu(DPM)$_3$] (see Tables 1 and 2) are in accord with the structures assigned to these germacranolides (Sanders and Williams, 1970; De Marco *et al.*, 1971). Zexbrevin B (**15a**) and another germacranolide, orizabin (**15b**), isolated from *Thitonia tubaeformis* Jacq Cass., possess very similar structures (Ortega *et al.*, 1971). They differ as in the case of calaxin (**13a**) and ciliarin (**13b**) in the ester group at C-8. Methacrylic acid esterifies the C-8 hydroxyl group of zexbrevin B (**15a**) and isobutyric acid the same function in orizabin (**15b**). **15a** and **15b** contain a secondary hydroxyl group substituted at C-2 and a hemiacetal function at C-3 whose ethereal oxygen is involved in a furan ring. Chromium trioxide oxidation of zexbrevin B (**15a**) and of orizabin (**15b**) gave calaxin (**13a**) and ciliarin (**13b**), respectively. The above correlation and those of calaxin (**13a**) and ciliarin (**13b**) with zexbrevin (**12**) show that these germacranolides obtained from different genera of the Compositae possess the same stereochemistry at C-6, C-7, C-8, and C-10.

TABLE 1

NUCLEAR MAGNETIC RESONANCE (NMR) SPECTRA OF ZEXBREVIN (**14**)[a]

Protons	Chemical shifts without Eu (DPM)$_3$[b]	Chemical shifts after addition of 25 mg of Eu (DPM)$_3$
C-4 CH_3	1.38	1.46
C-10 CH_3	1.39	1.63
Vinylic CH_3	1.85	2.02
H_6	4.50	4.80
H_8	5.20	5.56
H_2	5.56	5.89
Vinylic protons of the ester	5.60	5.63
group	5.98	6.41
Exocyclic methylene protons	5.70	5.82
	6.37	6.52

[a] The NMR spectra were determined by Eduardo Díaz in $CDCl_3$ on a Varian A60A spectrometer.

[b] Eu(DPM)$_3$ = tris(dipivalomethanate)europium (III).

TABLE 2

NUCLEAR MAGNETIC RESONANCE (NMR) SPECTRA OF CALAXIN (**25a**)[a]

Protons	Chemical shifts without Eu(DPM)$_3$[b]	Chemical shifts after addition of 25 mg of Eu(DPM)$_3$
C-10 CH_3	1.49	1.71
C-4 CH_3	1.87	1.91
Vinylic CH_3 of the ester group	2.09	2.13
H_9 and H_9'	2.43	2.78
H_7	3.73	3.95
H_6 and H_8	5.28	5.60
H_2	5.62	5.91
Vinylic protons of ester group	5.62	5.60
	6.05	6.32
Exocyclic methylene protons	5.71	5.80
	6.37	6.50
H_5	5.97	6.09

[a] The NMR spectra were determined by Eduardo Diaz in $CDCl_3$ on a Varian A60A spectrometer.
[b] Eu(DPM)$_3$ = tris(dipivalomethanate)europium(III).

The structure of zexbrevin C (**16**) has been established very recently in this laboratory. Although it does not contain a furan ring, it is closely related to the furan germacranolides described above.

Zexbrevin C (**16**) possesses a C-6-oriented five-membered lactone ring, a C-8 hydroxyl group esterified by methacrylic acid and a secondary hydroxyl group at C-1. A tertiary hydroxyl group at C-10 and a C-2 double bond could be the functional groups which serve as biogenetic precursors of the furan ring. Periodic acid or manganese dioxide oxidation of **16** gave the keto aldehyde with structure **17**.

Zexbrevin D possesses the C-1(10) and C-4 double bonds, a γ-lactone closed to C-6 and hydroxyl groups at C-8 and C-15, both esterified by acetic acid. Spectroscopic evidence combined with spin decoupling experiments led to the assignment of structure **18** for zexbrevin D.

It has been reported that *Artemisia mexicana* Willd. contains two guaianolides: estafiatin (**19**) (Sánchez-Viesca and Romo, 1963) and chrysartemin A (**20**) (Romo *et al.*, 1970) and two eudesmanolides: douglanin

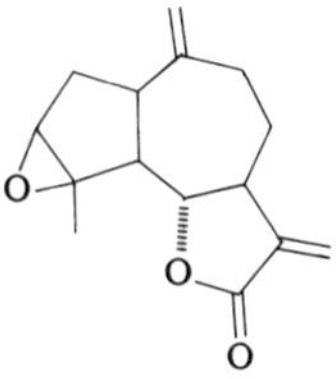

Estafiatin

19

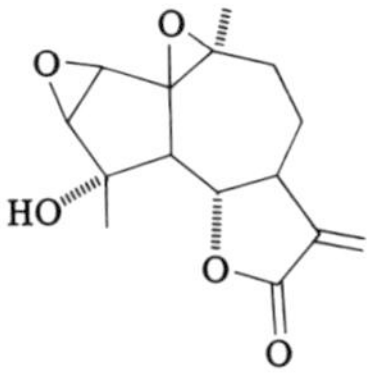

Chrysartemin A

20

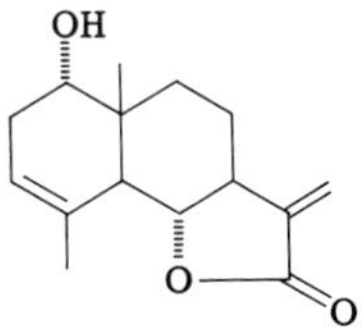

Douglanin

21

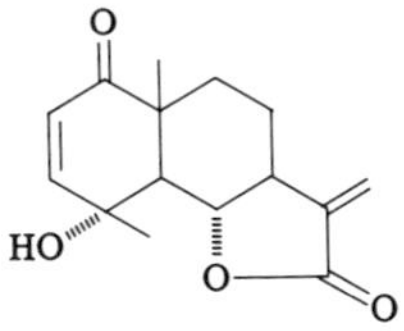

Arglanin

22

(**21**) (Matsueda and Geissman, 1967a) and arglanin (**22**) (Matsueda and Geissman, 1967b), isolated previously from *A. douglasiana* Bess.

We have recently found a germacranolide in *A. mexicana*, which proved to be identical to artemorin (**23**) a germacranolide isolated by Geissman

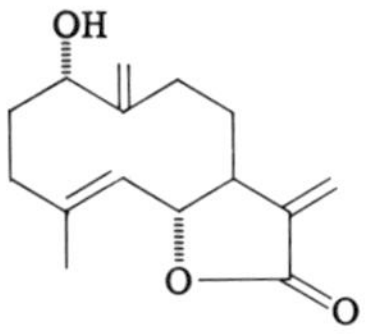

Artemorin

23

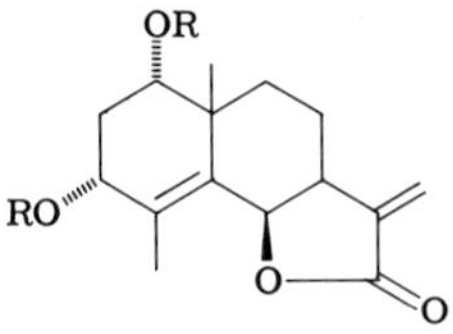

24a. R = H (Armexin)
b. R = Ac

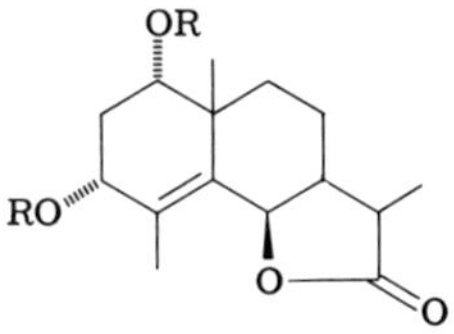

25a. R = H
b. R = Ac

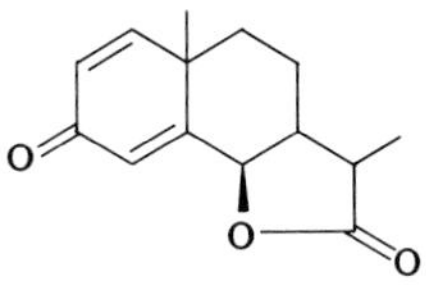

Isosantonin

26

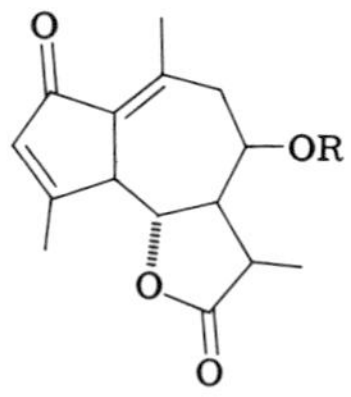

27a. R = H (Desacetyl-matricarin)
b. R = Ac (Matricarin)

and co-workers from *Artemisia verlotorum* Lamotte (Geissman and Lelo, 1971). This germacranolide appears to be the precursor of the other sesquiterpenes isolated from *A. mexicana* Willd.

A new eudesmanolide named armexin (**24a**) was isolated from *Artemisia mexicana* Willd. as the diacetate (**24b**). It differs from the other constituents isolated previously from this species in the stereochemistry at C-6. Armexin appears to be the first santanolide isolated in nature possessing a cis orientation of the lactone fusion. Catalytic hydrogenation of armexin diacetate (**24b**) hydrogenated only the methylene group conjugated with the γ-lactone. Alkaline hydrolysis of the dihydro derivative (**25b**), followed by manganese dioxide oxidation of the resulting diol (**25a**) gave isosantonin (**26**) (Barton *et al.*, 1962).

Desacetylmatricarin (**27a**), matricarin (**27b**), and chrysartemin A (**20**) have been isolated from several collections of *Artemisia klotzchiana* Bess. distributed in the arid regions of the central High Plateau of Mexico.

Acknowledgment

We are grateful to Syntex, S.A., for financial assistance.

References

Barton, D. H. R., and P. de Mayo. 1957. Sesquiterpenoids. Part VIII. The constitution of pyrethrosin. *J. Chem. Soc.* p. 150.

Barton, D. H. R., J. E. D. Levisalles, and J. T. Pinhey. 1962. Photochemical transformations. Part XIV. Some analogues of isophotosantonic lactone. *J. Chem. Soc.* p. 3472.

Benesová, V. A., Z. Samek, V. Herout, and F. Šorm. 1970. The structure of nobilin. *Tetrahedron Lett.* p. 5016.

De Marco, P. F., T. K. Elzey, R. B. Lewis, and E. Wenkert. 1971. Tris (dipivalomethanato) europium (III). A shift reagent for use in the proton magnetic resonance analysis of steroids and terpenoids. *J. Amer. Chem. Soc.* **92**:5737.

Geissman, T. A., and K. H. Lelo. 1971. Sesquiterpene lactones of *Artemisia*. Artemorin and dehydroartemorin (Anhydroverlotorin). *Phytochemistry* **10**:419.

Kupchan, S. M., Y. Ainehshi, J. M. Cassady, A. T. McPhail, G. A. Sim, H. K. Schnoes, and A. L. Burlingame. 1966. Structure elucidation of the novel sesquiterpenoids from *Elephantopus elatus*. *J. Amer. Chem. Soc.* **88**:3674.

Matsueda, S., and T. A. Geissman. 1967a. Sesquiterpene lactones of Artemisia species. IV. Douglanine from *Artemisia douglasiana* Bess. *Tetrahedron Lett.* p. 2159.

Matsueda, S., and T. A. Geissman. 1967b. Sesquiterpene lactones of Artemisia species. III. Arglanine from *Artemisia douglasiana* Bess. *Tetrahedron Lett.* p. 2013.

Ortega, A., A. Romo de Vivar, E. Díaz, and J. Romo. 1970. Determination de las estructuras de la calaxina y de la ciliarina, nuevos germacranólidos furanónicos. *Rev. Latinoamer. Quim.* **1**:81.

Ortega, A., C. Guerrero, A. Romo de Vivar, J. Romo, and A. Palafox. 1971. La orizabina y la zexbrevina B, nuevos germacranólidos furánicos. *Rev. Latinoamer. Quim.* **2**:38.

Rao, S., G. R. Kelkar, and S. C. Bhattacharyya. 1960. The structure of costunolide, a new sesquiterpene lactone from costus root oil. *Tetrahedron* **9**:275.

Rao, S., A. Sadgopal, and S. C. Bhattacharyya. 1961. Structure of saussurea lactone. *Tetrahedron* **13**:319.

Romo, J., A. Romo de Vivar, R. Treviño, P. Joseph-Nathan, and E. Díaz. 1970. Constituents of *Artemisia* and *Chrysanthemum* species. The structures of chrysartemins A and B. *Phytochemistry* **9**:1615.

Romo, J., A. Romo de Vivar, A. Ortega, E. Díaz, and M. A. Cariño. 1971. Las estructuras de la zinarosina y de la dihidrozinarosina, componentes de la *Zinnia acerosa* D.C. (Gray). *Rev. Latinoamer. Quim.* **2**:24.

Romo de Vivar A., A. Cabrera, A. Ortega, and J. Romo. 1967. Constituents of zaluzania species—II Structures of zaluzanin C and zaluzanin D. *Tetrahedron* **23**:3903.

Romo de Vivar, A., C. Guerrero, E. Díaz, and A. Ortega. 1970. Structure and stereochemistry of zexbrevin, a 3(2*H*) furanone germacranolide. *Tetrahedron* **26**:1657.

Sánchez-Viesca, F., and J. Romo. 1963. Estafiatin, a new sesquiterpene lactone isolated from *Artemisia mexicana* Willd. *Tetrahedron* **19**:1285.

Sanders, J. K. M., and D. H. Williams. 1970. A shift reagent for use in nuclear magnetic resonance spectroscopy. A first order spectrum of *n*-hexanol. *Chem. Commun.* p. 422.

Sousek, M., V. Herout, and F. Šorm. 1961. On terpenes CXVIII. Constitution of parthenolide. *Collect. Czech. Chem. Commun.* **26**:803.

Yoshioka, A., and T. J. Mabry. 1969. The structure and chemistry of isabelin. A new germacranolide dilactone from *Ambrosia psilostachya* D.C. compositae. *Tetrahedron* **25**:4767.

Yoshioka, H., T. J. Mabry, and E. Miller. 1968. Isabelin, a novel germacranolide dilactone from *Ambrosia psilostachya* D.C. *Chem. Commun.* p. 1679.

CHEMISTRY AND BIOGENESIS OF THE QUASSINOIDS (SIMAROUBOLIDES)

JUDITH POLONSKY

Institut de Chimie des Substances Naturelles, C.N.R.S., Gif-sur-Yvette, France

Introduction

The Simaroubaceae form a large botanical family of pantropical distribution; only a few representatives (such as *Ailanthus*) have extended into temperate regions. About fifteen of the simaroubaceous species are known to furnish bitter drugs, which have been used extensively in folk medicine. Indeed, they are employed by the local population to combat dysentery, fevers, amebiasis, and so on. At present, there are on the pharmaceutical market several medicinal preparations from the Simarouba plants which are recommended as amebicides.

Economically the Simaroubaceae family is important for various bitters prepared from the bitter principles present in the bark of most members. Some of them are also used as ornaments: the "tree of heaven" is nothing other than *Ailanthus altissima*.

Many species of the Simaroubaceae, particularly *Quassia amara*, have been known, for over a century, to contain bitter substances called collectively "quassin." But it was only in 1937 that the two major constituents of the wood *Quassia amara*, quassin and neoquassin, were isolated by Clark in a pure state, and an enormous number of chemical investigations have been carried out on them, especially by Robertson and his co-workers (Beer *et al.*, 1956). However, the elucidation of their structures was not performed until modern physical techniques of investigation were available. Between 1960 and 1962, Valenta and his co-workers, making extensive use of nuclear magnetic resonance (NMR) data and incorporating the earlier results of modern physical techniques of investigation were available. Between 1960 and 1962, Valenta and his co-workers, making extensive use of nuclear magnetic resonance (NMR) data and incorporating the earlier results of Robertson and his school, determined the structure of quassin (**1**) and neoquassin (**2**). These are thus the first bitter principles whose structures have been fully elucidated. It can be seen that quassin and neoquassin are dimethoxy derivatives of C-20 compounds: quassin is a δ-lactone and neoquassin is the corresponding hemiketal. They can be converted into one another: oxidation of neoquassin gives quassin, and reduction of the latter yields neoquassin. As will be seen later, the easy reduction of this δ-lactone is found again in other bitter simaroubaceous constituents.

By establishing the structure of quassin, Valenta and his co-workers thus resolved one of the classical problems of plant structural chemistry. It is interesting to note that another problem which has engaged the interest and ingenuity of natural product chemists for well over a century was solved in the same year—the structure of limonin (**3**), the characteristic bitter principle of the *Citrus* species of the Rutaceae family. A great number of compounds isolated from the same or a botanically related family, the Meliaceae, have a structure similar to that of limonin, and they form the bitter principles of the limonin group, called limonoids (Connoly *et al.*, 1970). Those isolated from the Meliaceae family, also called meliacins, have ring A intact; this is exemplified by gedunin (**4**).

The simaroubaceous bitter constituents share the same biosynthetic pathway with limonin and its relatives.

To return to the bitters of the Simarubaceae, since the early 1960's, when the structures of quassin and neoquassin were established, a number of genera of the family have been investigated. Many new bitter simarouba-

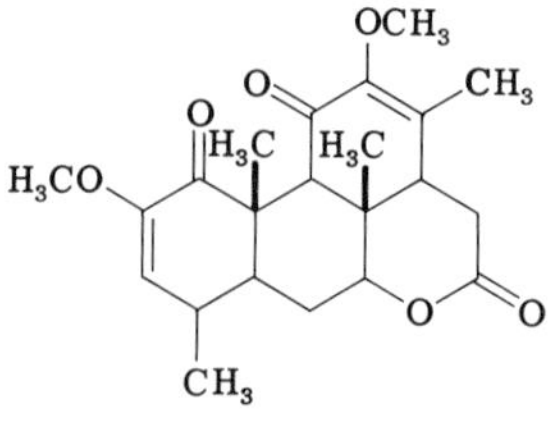

OCH$_3$ O CH$_3$ O H$_3$C H$_3$C H$_3$CO O O CH$_3$

1

OCH$_3$ O CH$_3$ O H$_3$C H$_3$C H$_3$CO OH O H CH$_3$

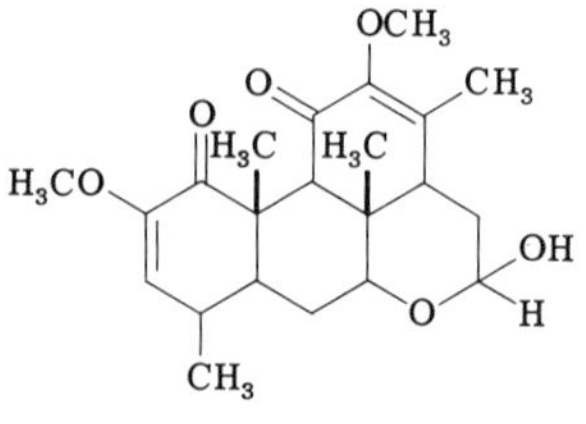

2

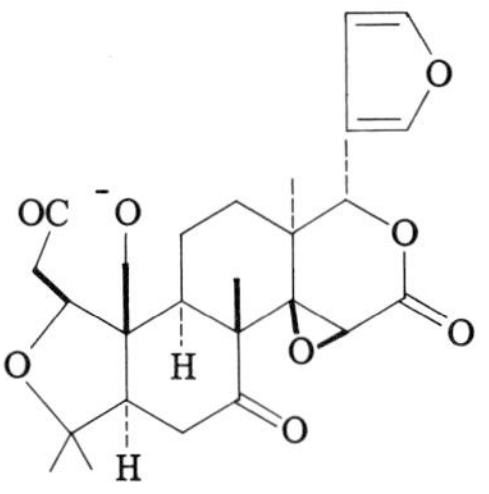

3 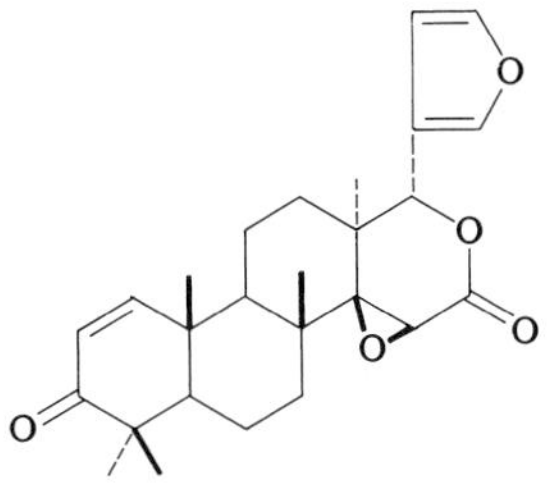 4

ceous constituents have been isolated, and their structures have been elucidated (Connoly *et al.*, 1970). These new bitter principles, whether they have, fundamentally, 19, 20, or 25 carbon atoms, are all closely related chemically, they form a new family of compounds called—because of their relationship to quassin—quassinoids. Sometimes they are also designated by the name simaroubolides.

Isolation and Structure of Various Simaroubaceae Bitter Constituent

We have investigated eleven different simaroubaceous species; these are listed in Table 1. In the first column are the names of the species examined; in the second, the names of the bitter compounds with their molecular formulas.

We have examined either the bark or the seeds of these species, depending on their availability. For the extraction of the bitters we followed various procedures, all of which are rather tedious. In general, we began by defatting the ground and dried plant material by extraction with petroleum ether. This was followed by water extraction; although the pure compounds are relatively water insoluble, when impurities are present the whole extract is water soluble. The bitter compounds were isolated from the aqueous solution by continuous extraction with appropriate solvents. Repeated chromatography of the crude extract allowed the separation of the different bitter compounds. Thin-layer chromatography techniques were of great help in the determination of the purity.

Table 2 lists, along with *Quassia amara* and *Castela nicholsoni* (Hollands *et al.*, 1965; Geissman and Ellestad, 1962; Stöcklin and Geissman, 1970; Mitchell *et al.*, 1971), several other simaroubaceous species which have been studied recently by other workers. These are *Holocantha emoryi* (Stöcklin *et al.*, 1969; Wall and Wani, 1970), *Brucea sumatrana* (Sim *et al.*, 1968; Stöcklin and Geissman, 1968; Duncan and Henderson, 1968), *Picrasma ailanthoides* (=*P. quassioides*) (Hikino *et al.*, 1970a,b; Murae *et al.*,

TABLE 1

MOLECULAR FORMULAS OF COMPOUNDS DERIVED FROM SIMAROUBACEAE

Species	Compounds		
Simarouba amara		Simarolide $C_{27}H_{36}O_9$	
Simarouba glauca	Glaucarubin $C_{25}H_{36}O_{10}$	Glaucarubinone $C_{25}H_{34}O_{10}$	Glaucarubolone $C_{20}H_{26}O_8$
Perriera madagascariensis	Glaucarubin $C_{25}H_{36}O_{10}$	Glaucarubinone $C_{25}H_{34}O_{10}$	2-Acetylglaucarubin-one $C_{27}H_{36}O_{11}$
Hannoa klaineana	Klaineanone $C_{20}H_{28}O_6$	Chaparrinone $C_{20}H_{26}O_7$	Glaucarubolone $C_{20}H_{26}O_8$
Ailanthus altissima (=*Ailanthus glandulosa*)	Amarolide $C_{20}H_{28}O_6$	Chapparinone $C_{20}H_{26}O_7$	Ailanthinone $C_{25}H_{34}O_9$
	Acetyl amarolide $C_{22}H_{30}O_7$	Chaparrin $C_{20}H_{28}O_7$	Ailanthone $C_{20}H_{24}O_7$
Ailanthus grandisisma			Ailanthone $C_{20}H_{24}O_7$
Brucea amarissima	Brucein A $C_{26}H_{34}O_{11}$	Brucein B $C_{23}H_{28}O_{11}$	Brucein C $C_{28}H_{36}O_{12}$
	Brucein D $C_{20}H_{26}O_9$	Brucein E $C_{20}H_{28}O_9$	Brucein F $C_{20}H_{28}O_{10}$
Soulamea pancheri	Picrasin B $C_{21}H_{28}O_6$	Hydroxy-6-picrasin B $C_{21}H_{28}O_7$	
Quassia africana	Simalikalactone A (=nigakilactone B) $C_{22}H_{32}O_6$	Simalikalactone B (=picrasin B) $C_{21}H_{28}O_6$	Simalikalactone C (=dehydro-12-nigkilactone A) $C_{21}H_{28}O_6$
	Simalikalactone D $C_{25}H_{34}O_9$	Simalikahemiacetal A (=neoquassin) $C_{22}H_{30}O_6$	
Samadera indica (=*Samadera madagascariensis*)	Samaderin B $C_{19}H_{22}O_7$	Samaderin C $C_{19}H_{24}O_7$	Samaderin D $C_{19}H_{22}O_7$
Simaba cedron	Cedronin $C_{19}H_{24}O_7$	Cedronolin $C_{19}H_{26}O_7$	

1970a,b,c, 1971), and *Eurycoma longifolia* (Le-Van-Thoi and Nguyen-Ngoc-Suong, 1970).

Tables 1 and 2 include more than forty different compounds, whose molecular formulas vary to a rather large extent; their molecular weights lie between 360 and 560. All these bitter substances, which occur either as esters or in the free state, are closely related chemically. However, they can be divided into distinct groups according to their basic skeleton. Three such skeletons have been discerned: **5**, with 19 carbon atoms; **6**, with 20 carbon atoms; and **7**, with 25 carbons.

Only one representative of this last group (**7**, with 25 carbon atoms) was known before 1970. Since then a second has been isolated by Japanese workers. The majority of the simaroubaceous bitter principles belong to the C_{20} group (**6**), and they all possess a δ-lactone group. The compounds with 19 carbon atoms have lost one carbon atom from position 16, and they all have a γ-lactone function. The compounds with 25 carbons have an additional 5-carbon unit attached to position 13 and possess one δ- and one γ-lactone; the latter is reminiscent of the furan ring present in the limonoids.

TABLE 2

Molecular Formulas of Recently Studied Compounds Derived from Simaroubaceae

Species		Compounds	
Quassia amara	Quassin $C_{22}H_{28}O_6$	Neoquassin $C_{22}H_{30}O_6$	
Castela nicholsoni	Chaparrin $C_{20}H_{28}O_7$	Glaucarubol $C_{20}H_{28}O_8$	Glaucarubolone $C_{20}H_{26}O_8$
	Isovaleryl-15 glaucarubol $C_{25}H_{36}O_9$	Chaparrolide $C_{20}H_{30}O_6$	Castelanolide $C_{20}H_{28}O_6$
		Amarolide	
Holocantha emoryi	Glaucarubol $C_{20}H_{28}O_8$	Acetoxy-15 glaucarubolone (holocanthone) $C_{22}H_{28}O_9$	
Brucea sumatrana	Brusatol (dehydro-brucein A) $C_{26}H_{32}O_{11}$	Brucein D $C_{20}H_{26}O_9$	Brucein E (WST-63) $C_{20}H_{28}O_9$
	Brucein G $C_{20}H_{26}O_8$		
Picrasma ailanthoides (=*P.quassioides*)	Picrasin A $C_{26}H_{34}O_8$	Picrasin B $C_{21}H_{28}O_6$	
	Nigakilactone A $C_{21}H_{30}O_6$	Nigakilactone B $C_{22}H_{32}O_6$	Nigakilactone C $C_{24}H_{34}O_7$
	Nigakilactone D (=quassin)	Nigakilactone E $C_{24}H_{34}O_8$	Nigakilactone F $C_{22}H_{32}O_7$
	Nigakilactone G (=picrasin A)	Nigakilactone H $C_{22}H_{32}O_8$	Nigakilactone J $C_{23}H_{34}O_7$
	Nigakihemiacetal A $C_{22}H_{34}O_7$	Nigahihemiacetal B (=neoquassin)	
Eurycoma longifolia	Eurycomalactone $C_{19}H_{24}O_6$		

The number and the position of the methyl groups are the same on these three basic skeletons (**5**, **6**, and **7**). All the members of these groups so far known have only one methyl group at C-4.

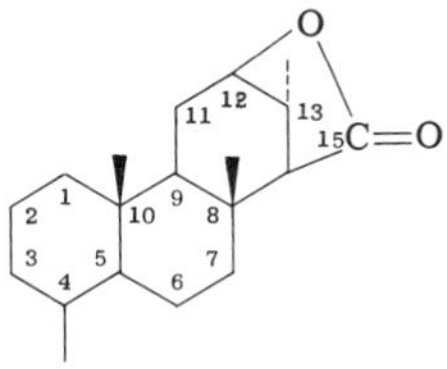

5

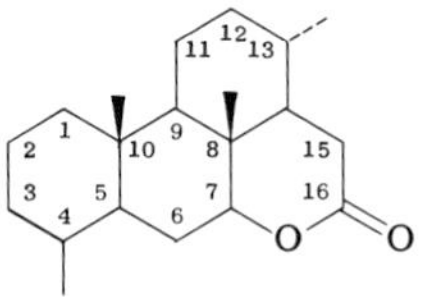

6

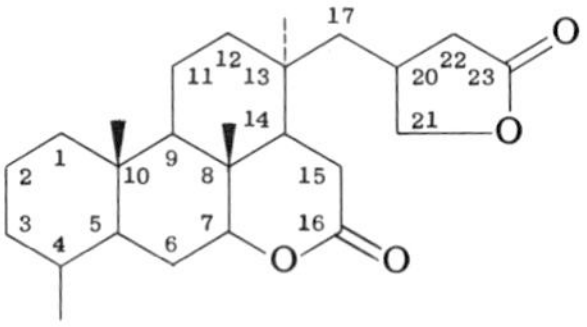

7

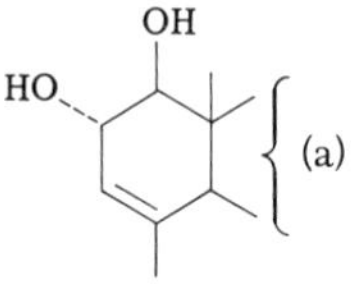

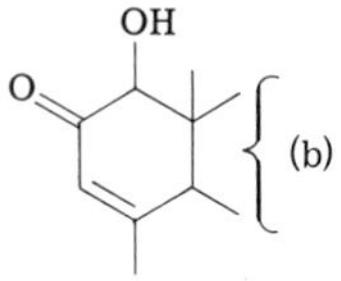

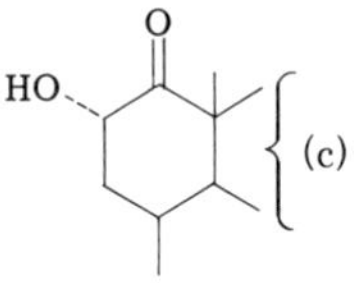

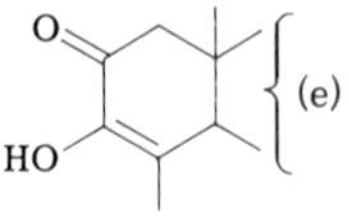

(a) (b) (c)

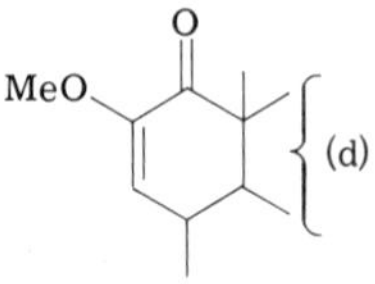

(d) (e)

The quassinoids are heavily oxygenated lactones (δ-lactone in the C_{20} compounds and γ-lactone in the C_{19} compounds) possessing rarely more than one double bond. They have varying numbers of oxygen-containing groups (for example, hydroxyl or esterified hydroxyl, carbonyl, oxide, methoxyl, or carboxymethyl). In the C_{20} compounds, these oxygenated functions may be found at most of the carbon atoms, although the methyl groups at C-4 and C-10 and the carbons C-5 and C-9 have never yet been found to have an oxygen function. Despite this widespread pattern of oxygenation, certain privileged positions emerge. Thus, all the quassinoids, except five, have an oxygenated group at positions 1, 2, 7, 11, and 12. The exceptions are the two C_{25} compounds, simarolide (**10**) and picrasin A (**11**), which possess no oxygenated function at position 12, and bruceins A, B, C and brusatol, which have oxygenated functions at positions 2 and 3 but not at 1 and 2.

The variations in the structures of quassinoides are principally as follows: Ring A may have the structures (a), (b), (c), (d), or (e). However, structure (e) has so far been found in only four quassinoids. Ring C may possess at position 8 either a methyl group or a hydroxymethyl which forms a hemiketal to C-11 or an oxide to C-13. Ring D may have a hydroxyl group at C-15, which is generally found esterified with various small fatty acids (acetic, 2-methylbutyric, isovaleric, senecioic, 2-hydroxy-2-methylbutyric or 3,4-dimethyl-4-hydroxyvaleric acid).

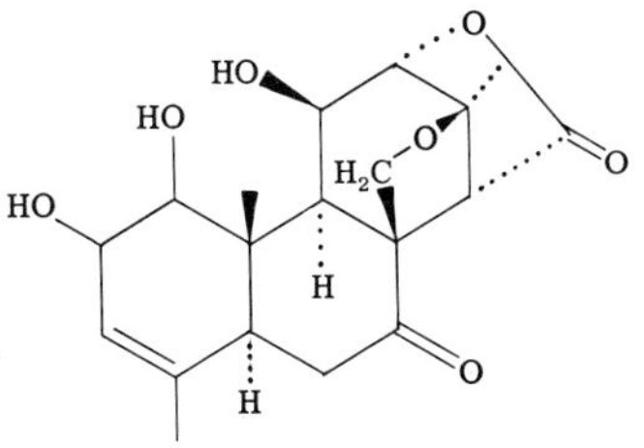

Samaderin C

8

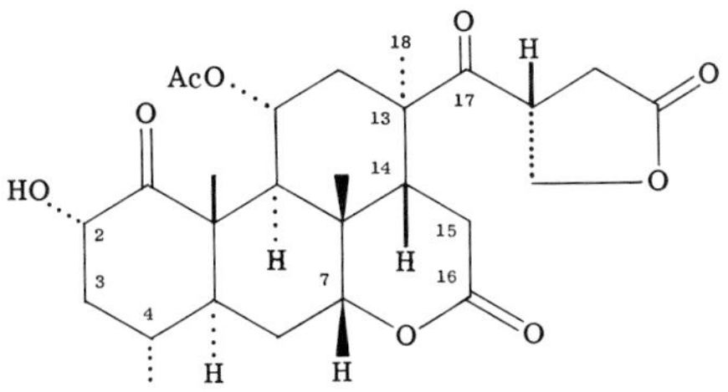

Simarolide

10

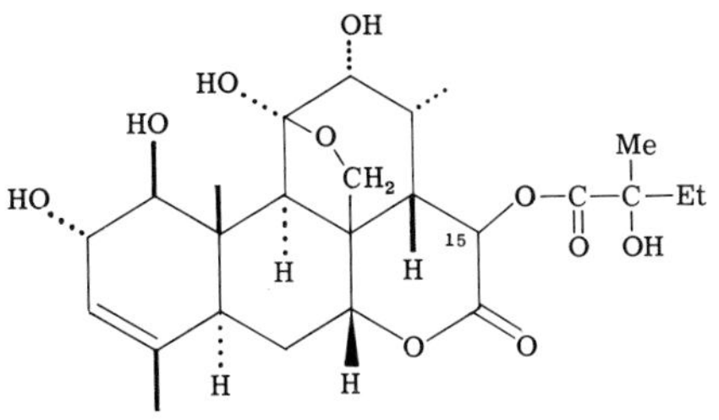

Glaucarubin

9

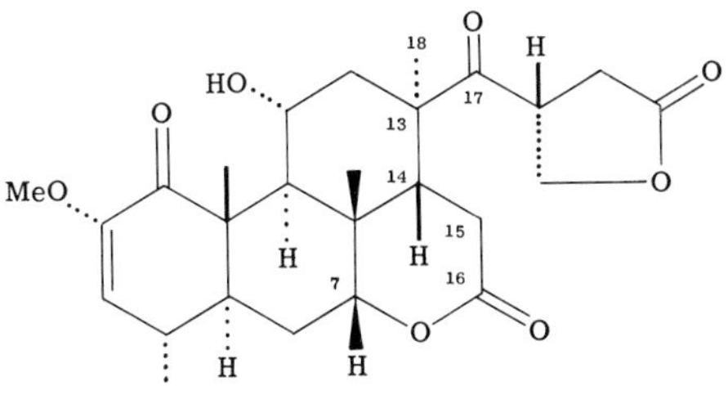

Picrasin A

11

Samaderin C (**8**) (Zylber and Polonsky, 1964), glaucarubin (**9**) (Polonsky *et al.*, 1964), and simarolide (**10**) (Polonsky, 1964; Brown and Sim, 1964) are representative examples of the three basic skeletons described above. All three have oxygenated functions at C-1 and C-2. Simarolide has a methyl at C-8, whereas in the other two, there is a hydroxy methyl involved in an epoxide linked to carbon-13 in samaderin and to carbon-11 in glaucarubin; in the latter case this function can be called a hemiacetal.

Picrasin A (**11**) is the second C_{25} quassinoid isolated quite recently by Hikino *et al.* (1970b); it has been correlated to simarolide.

Most of our work has been done on the three compounds **8**, **9**, and **10**, and understanding their chemical and physical properties was of great help in the structural studies of the other simaroubaceous bitter principles. Before describing some chemical and phsyical properties that are common to a number of them, I propose to review briefly the simaroubaceous species with their bitter principles that have been studied.

Glaucarubin (**9**) and samaderin C (**8**) have an α-glycol grouping adjacent to a double bond in ring A while other bitter substances possess an α-ketol instead of the diol system. It is the allylic hydroxyl which has been oxidized,

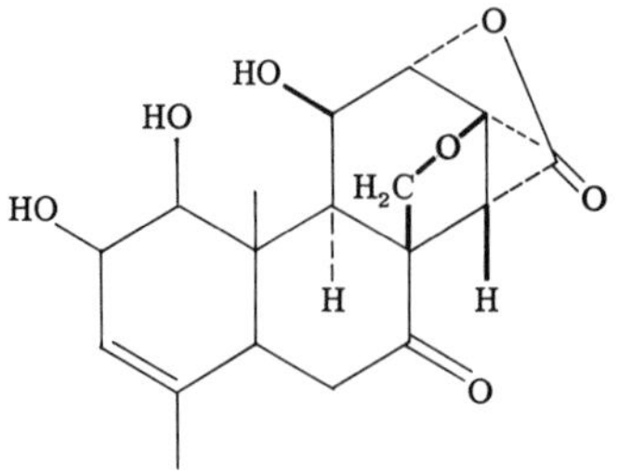

Samaderin C

8

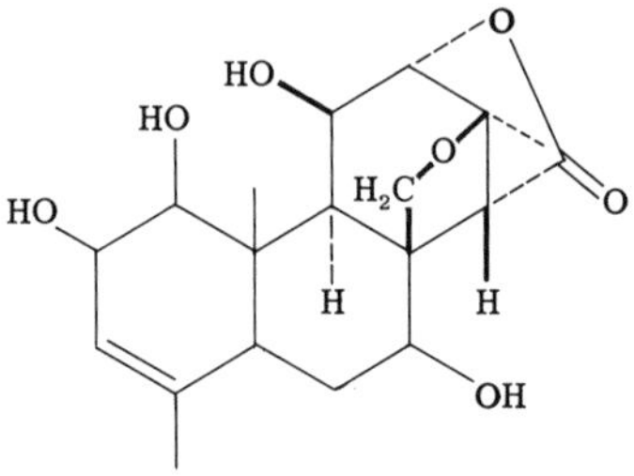

Cedronolin

13

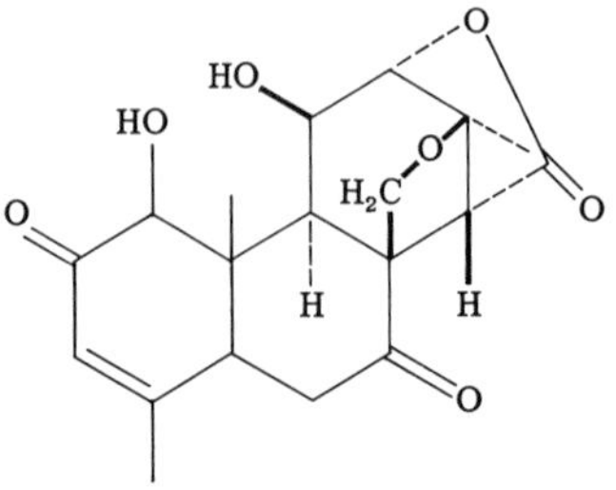

Samaderin B

12

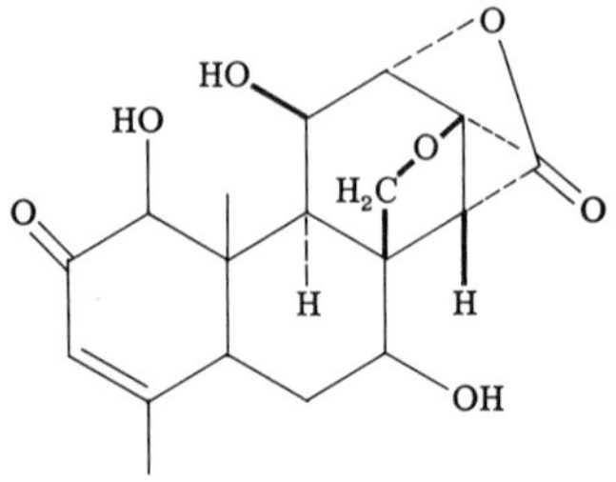

Cedronin

14

so that these compounds all show the characteristic ultraviolet absorption of an α,β-unsaturated ketone. Often, both types are found together in the same plant. Thus, a second bitter compound called samaderin B (**12**) (Zylber and Polonsky, 1964), which possesses an α-ketol instead of the α-glycol grouping, has been isolated along with samaderin C from *Samadera indica.*

From the fruit of *Simaba cedron*, cedronolin (**13**) together with cedronin (**14**) have been isolated (Zylber and Polonsky, 1964). These two compounds are dihydro derivatives of the samaderins, having a hydroxyl instead of the ketone at position 7.

In *Simaruba glauca*, the major constituent (glaucarubin) is accompanied by glaucarubinone (**15**) (Gaudemer and Polonsky, 1964) with the α-ketol

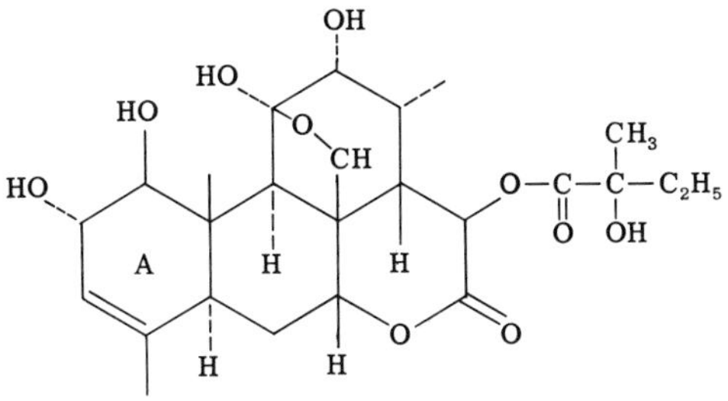

Glaucarubin

9

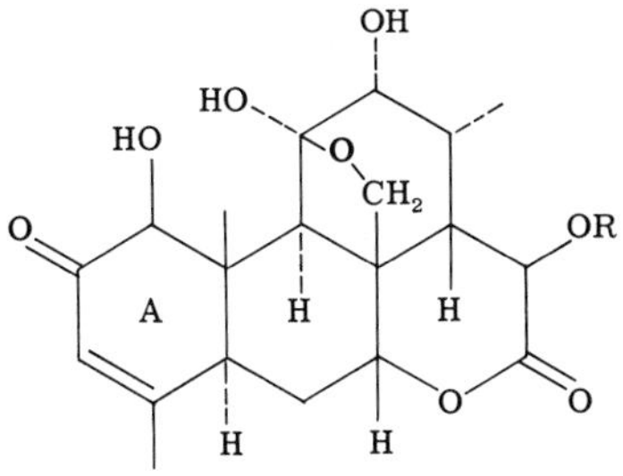

15. Glaucarubinone

$R = CO{-}C(CH_3)(OH){-}C_2H_5$

16. Glaucarubolone

R = H

grouping in ring A. We have also isolated a much smaller quantity of glaucarubolone (**16**) which has a free hydroxyl at C-15 and can easily be obtained by saponification of glaucarubinone.

Oxidation of the compounds of the first type with MnO_2 leads readily to those of the second type, that is to say, to those with the α,β-unsaturated ketone. On the other hand, the α-ketol can be reduced with $NaBH_4$ to the α-glycol grouping; but it should be noted that during this reaction the lactone is also reduced, leading to a hemiketal.

It is of interest that the facile reduction of the lactone group with sodium borohydride, and to a lesser extent by catalytic hydrogenation, has proved to be a characteristic feature of all the quassinoids having a six-membered lactone ring.

In glaucarubin and glaucarubinone, the hydroxyl at C-15 is esterified with α-methyl-α-hydroxybutyric acid. However, in other principles this hydroxyl group is found esterified with other acids: for instance, acetic

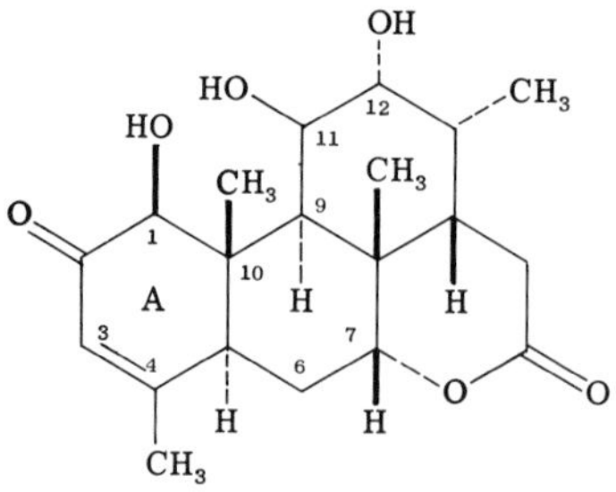

Klaineanone

19

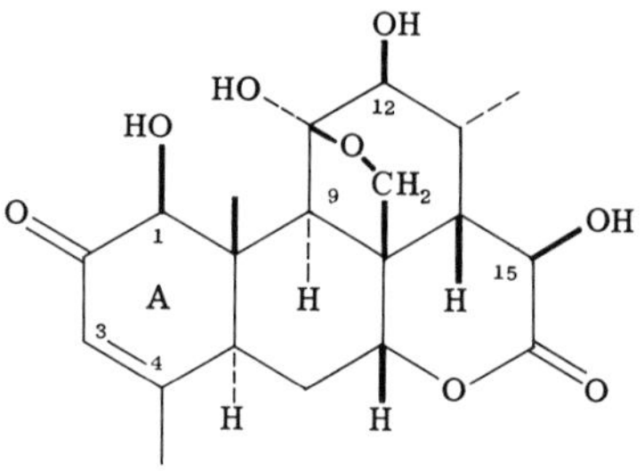

Glaucarubolone

16

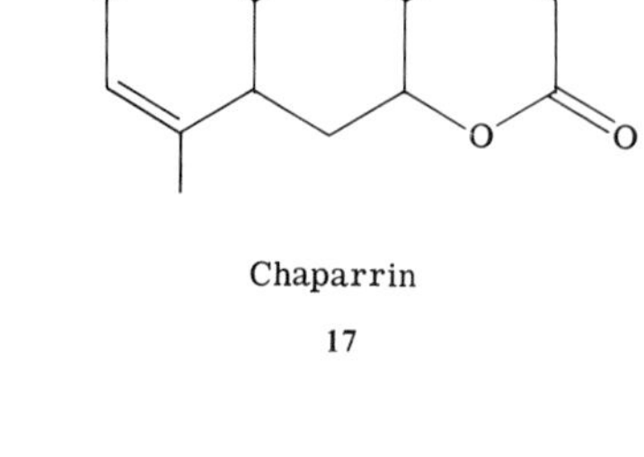

Chaparrin

17

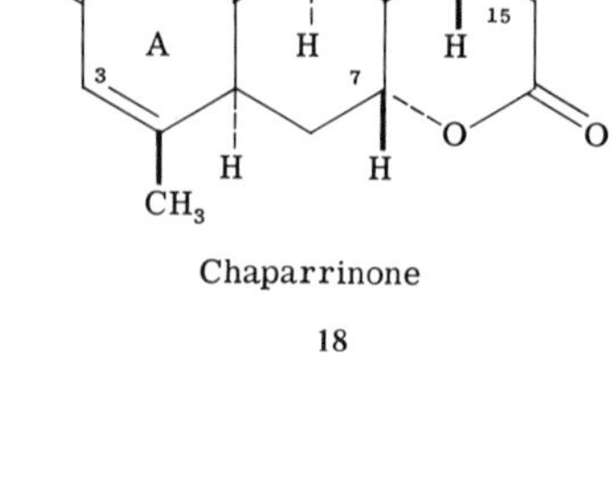

Chaparrinone

18

acid, isovalerianic acid, isobutyric acid, senecioic acid. This variety makes the isolation and separation of the bitter principles more tedious. Several bitter principles do not have a hydroxyl at C-15, such as chaparrin (**17**) (Geissman and Ellestad, 1962; Hollands *et al.*, 1965) which is the 15-deoxyglaucarubol.

Hannoa klaineana also contains bitter substances without a hydroxyl at C-15. From the nuts of this plant three bitter compounds have been isolated (Polonsky and Bourguignon-Zylber, 1965). One of them was readily identified as glaucarubolone (**16**), which is the compound obtained by alkaline hydrolysis of glaucarubinone (**15**). The second substance isolated in much greater quantity differs from glaucarubolone by the absence of the hydroxyl at C-15. It is therefore a 2-dehydrochaparrin and has been named chaparrinone (**18**). The third compound, called klaineanone (**19**), is the least oxygenated compound, containing only 6 oxygen atoms. It possesses a methyl group at C-8. Here we have an example of a simaroubaceous plant which synthesizes three bitter principles having different states of oxidation.

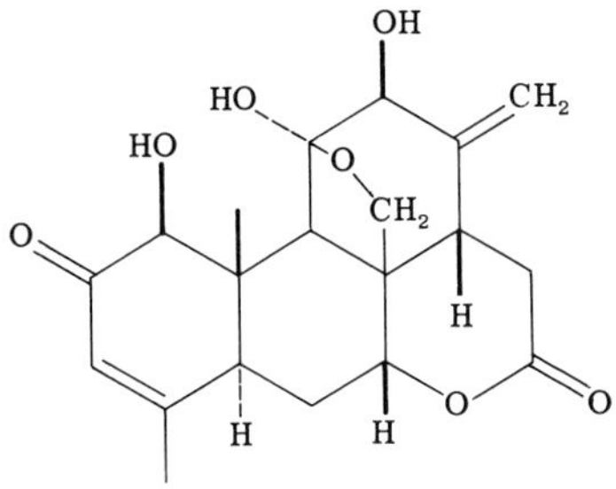

Ailanthone

20

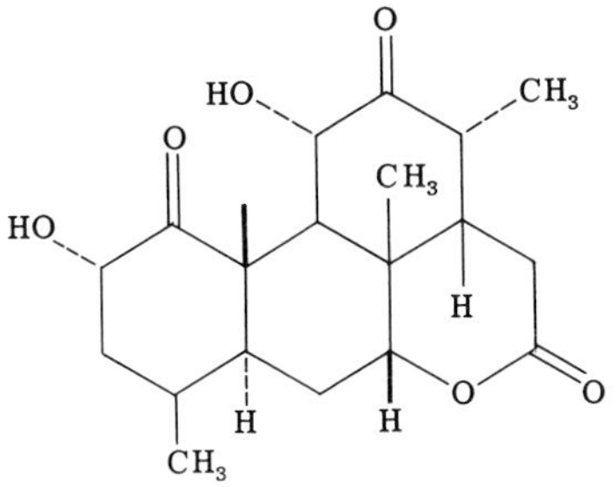

Amarolide

21

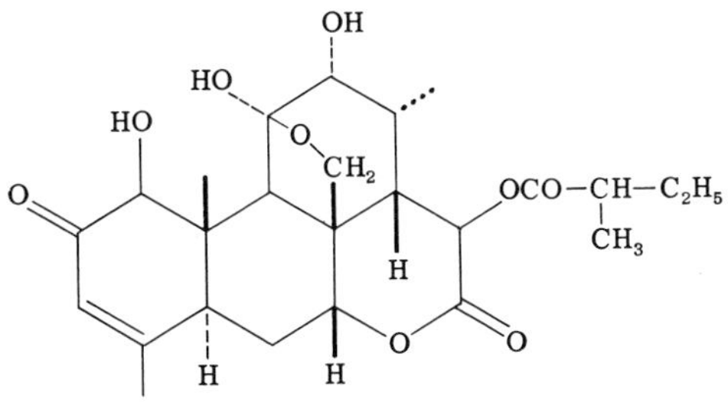

Ailanthinone

22

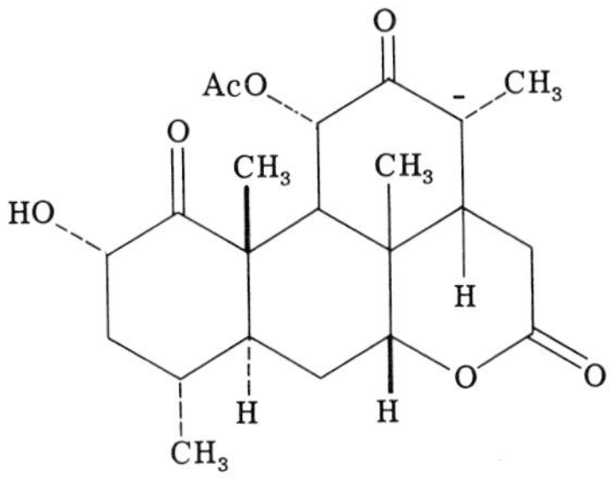

Amarolide acetate

23

Ailanthus altissima synthesizes a multitude of bitter principles. Several of them have been isolated in the pure state. The major constituent is ailanthone (**20**) (Polonsky and Fourrey, 1964). It reveals two new structural features; in contrast to the other bitter principles, the hydroxyl at C-12 in ailanthone is not α-axial but β-equatorial; furthermore it does not have the secondary methyl group at C-13, but possesses an exomethylene.

As minor constituents, amarolide (**21**) and its acetate (**23**) as well as ailanthinone (**22**), which possesses a methyl butyric ester function at C-15, have been extracted (Fourrey, 1968). Ailanthone, amarolide, and its acetate have also been isolated by Casinovi *et al.* (1964; 1965) from the roots of *Ailanthus glandulosa*, which is synonymous with *Ailanthus altissima*. The structure of amarolide, which has been related to quassin, has been established by the Italian authors and by Stöcklin *et al.* (1970).

From the seeds of *Brucea amarissima* may be extracted a complex mixture of compounds. Five of these have been isolated in the pure state. We have called them bruceins A to F.

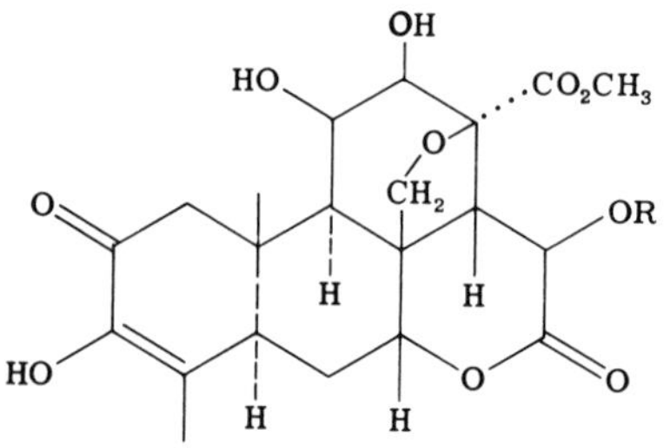

24. Brucein A: $C_{26}H_{34}O_{11}$: R = $-CO-CH_2-CH(CH_3)_2$

25. Brucein B: $C_{23}H_{28}O_{11}$: R = $-CO-CH_3$

26. Brucein C: $C_{28}H_{36}O_{12}$: R = $-CO-C(CH_3)=CH-C(OH)(CH_3)_2$

Bruceins A, B, and C, (**24**), (**25**), and (**26**) (Polonsky *et al.*, 1967), differ one from another by the nature of the acid that esterifies the hydroxyl at C-15. They are, respectively, isovalerianic, acetic, and 2,4-dimethyl-4-hydroxy-2-pentenoic acid. In brusatol (Sim *et al.*, 1968) the hydroxyl at C-15 is esterified with senecioic acid.

Bruceins A, B, and C and brusatol are the most oxygenated bitter compounds so far isolated from the Simaroubaceae. They reveal new structural features: they possess a carbomethoxy gorup at C-13,and, in contrast to all the other simaroubaceous bitter principles, the oxygenated functions in ring A are not found at positions 1 and 2, but at 2 and 3.

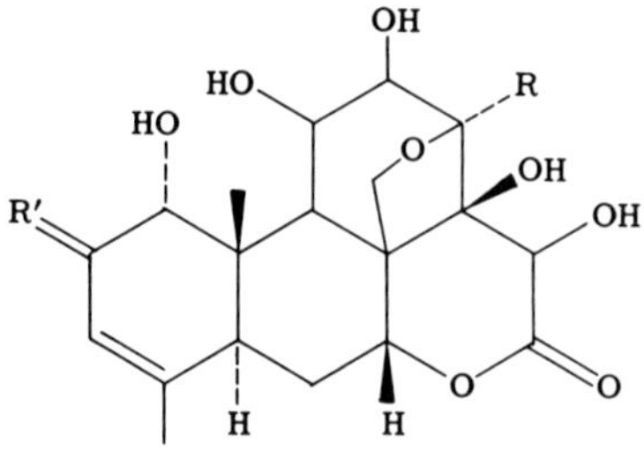

27. Brucein D: R′ = O; R = CH_3: $C_{20}H_{26}O_9$
28. Brucein E: R′ = H; OH R = CH_3: $C_{20}H_{28}O_9$
29. Brucein F: R′ = H; OH R = CH_2OH: $C_{20}H_{28}O_{10}$

In bruceins D (**27**), E (**28**) (Polonsky *et al.*, 1968), and F (**29**) (Polonsky *et al.*, 1969), we find again the oxygenated functions at C-1 and C-2. The co-occurrence of both types of compounds in the same plant may have some biogenetic significance. Actually, they may be supposed to arise from a unique precursor having an oxygenated function at position 3. The formation of 1,2-oxygenated compounds from a 3-oxygenated compound may be formally explained by several schemes, one of which is the following:

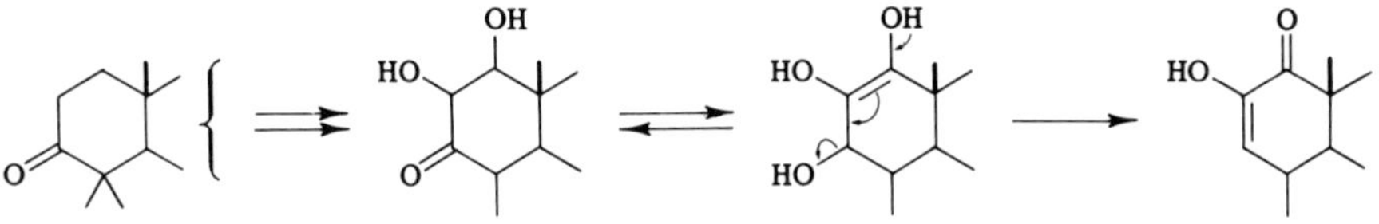

(c) (d)

The latter transformation (c⟶ d) has its analogy in the Mattox rearrangement (Mattox, 1952).

Bruceins D, E, and F are the first bitter principles to possess an α-glycol grouping at C-14–C-15, and this may also be significant from the biogenetic

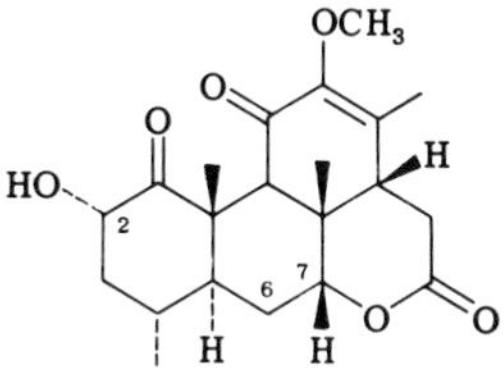

Picrasin B

30

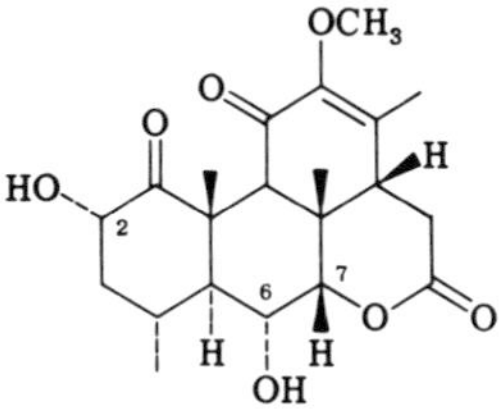

6-Hydroxypicrasin B

31

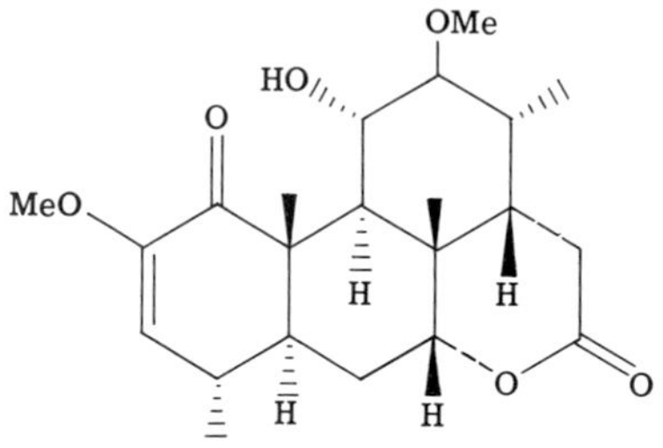

Simalikalactone A
(= Nigakilactone B)

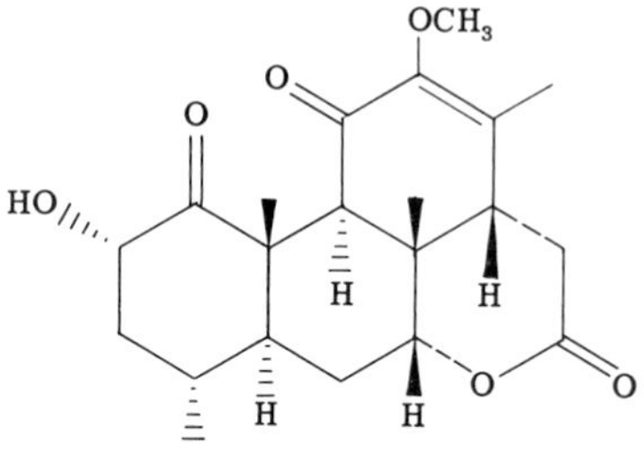

Simalikalactone B
(= Picrasin B)

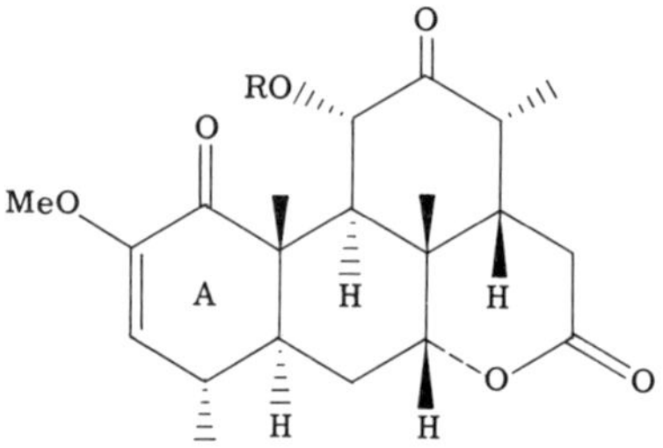

Simalikalactone C

32

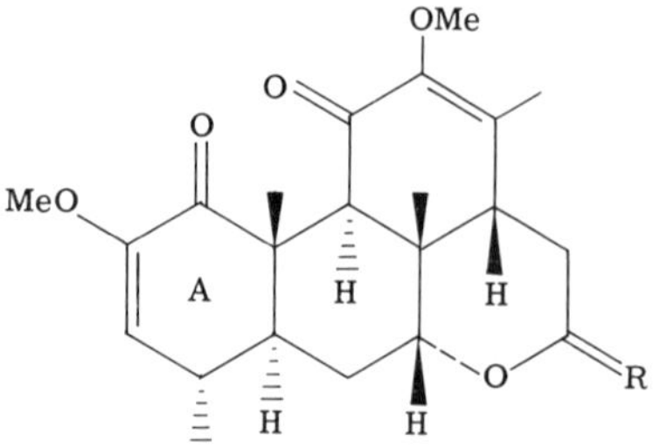

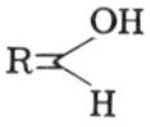

R= $\langle^{OH}_{H}$: Simalikahemiacetal
(= Neoquassin)

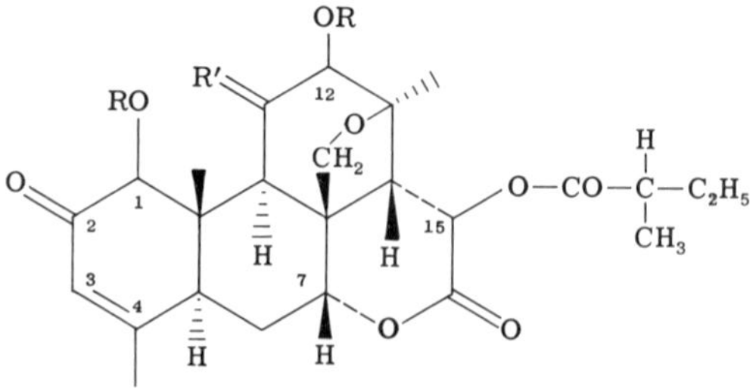

$R = H$

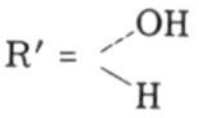

$R' = \langle^{OH}_{H}$

Simalikalactone D

33

point of view. This grouping may be formed by hydroxylation of the double bond C-14–C-15 of apoeuphol, the supposed biogenetic precursor of the quassinoids, as will be seen later.

One of the most recently examined simaroubaceous plants is *Soulamea pancheri* (Viala and Polonsky, 1970), originating from New Caledonia. Two quassinoids have been isolated. The first turned out to be identical with picrasin B (**30**), a bitter principle isolated quite recently by Hikino *et al.* (1970a) from another simaroubaceous plant; we have correleated it to quassin by oxidation with bismuth oxide followed by methylation. To the second quassinoid we were able to attribute the structure of 6-hydroxypicrasin (**31**). It is one of the rare quassinoids having a substituent at position 6.

Finally, the last simaroubaceous species studied by us is *Quassia africana* B (Tresca *et al.*, 1971); this originated in the Congo, where it is named "Simalikali" (meaning "more bitter than anything"). We have isolated five quassinoids, called simalikalactones and simalikahemiacetal. Three of them have been identified as known compounds. The two others, simalikalactone C and simalikalactone D have the structures **32** and **33**. The latter is the α-methyl butyryl ester of 14-deoxybrucein D. Here we have an example of a simaroubaceous species which synthesizes quassinoids of two different types.

Picrasma ailanthoides P has been studied by Murae *et al.* (1970a,b,c, 1971). Seven quassinoids, named nigakilactones A to J, have been isolated. Three of them have been identified with known compounds, the others have been correlated to quassin. Three hemiacetals have been isolated, one of which was identical to neoquassin.

We can thus see the large diversity of modifications that have been observed on the basic skeletons of the simaroubaceous bitter principles.

In contrast to such groups as alkaloids and glucosides, there are no useful diagnostic tests for the rapid recognition of the presence of these bitter substances in plants besides the bitter taste. It has been claimed in the literature that the simaroubaceous bitter principles give a color reaction with concentrated sulfuric acid. However, this reaction is far from general and seems to be dependent on the structure of ring A.

Chemical and Physical Properties of the Quassinoids

Certain chemical and physical properties are common to a number of the quassinoids.

1. All the quassinoids which have an α-glycol grouping and a double bond in ring A, such as glaucarubol (**34**) (the alkaline hydrolysis product of

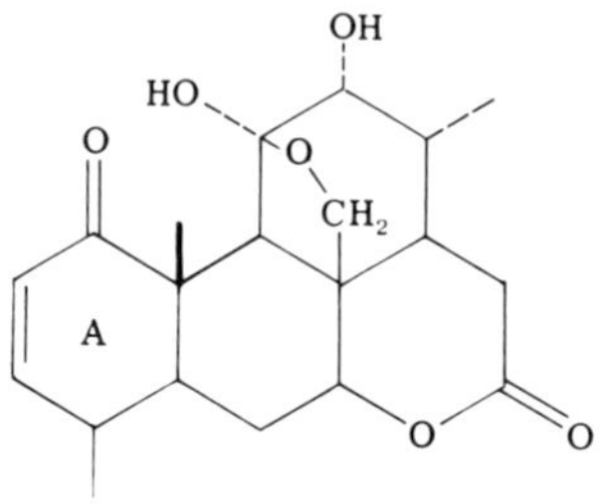

36

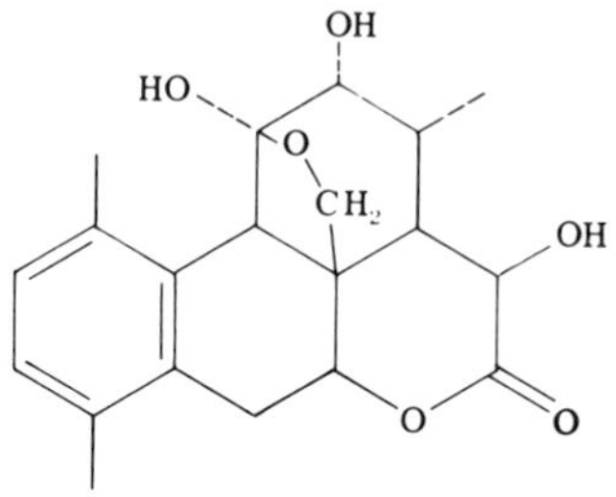

Glaucanol

35

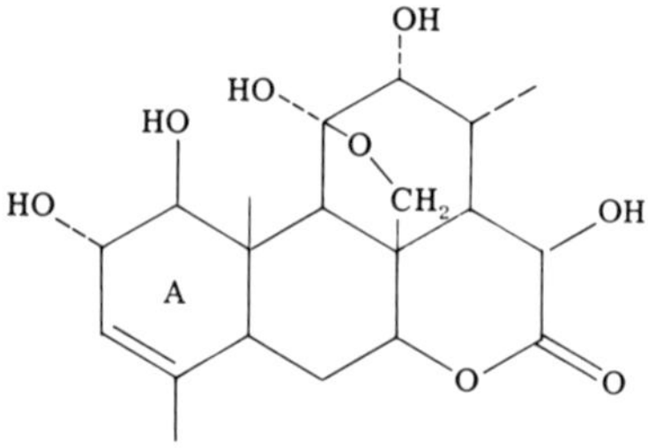

Glaucarubol

34

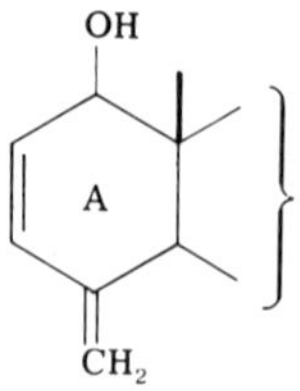

37

glaucarubin), undergo readily an acid-catalyzed dehydration leading finally to two compounds having the structure of ring A, alone, modified. The major product has its A ring aromatized as glaucanol (**35**), and the minor one is an unsaturated ketone (**36**).

This dehydration has been shown to take place in two stages. There is at first loss of one molecule of water to give dienes with structural formula **37**, which have been isolated as crystals from several quassinoids by reaction with very mild acids. On subsequent treatment with mineral acids, the dienes may be induced to isomerize to the unsaturated ketones (**36**) as well as to lose a second molecule of water leading to the benzenoid A ring. The latter conversion, **37** → **35**, is essentially a dienol-benzene rearrangement.

2. A number of quassinoids possess a hemiketal in ring C involving the carbonyl function at C-11 and the hydroxymethyl at C-8. In the absence of additional carbonyl groups, these compounds do not have a UV absorption associated with saturated ketones, not do they show a Cotton effect in their optical rotatory dispersion or dichroism curves. This preferred hemiketal formation can be clearly rationalized by the stereochemical

conformation of this type of compounds. Models show that, with the A/B and B/C ring junctions *trans* fused, the molecule is well oriented for hemiketal formation, and such formation decreases the nonbonded interaction between the α-axial hydroxyl at C-12 and carbon C-15.

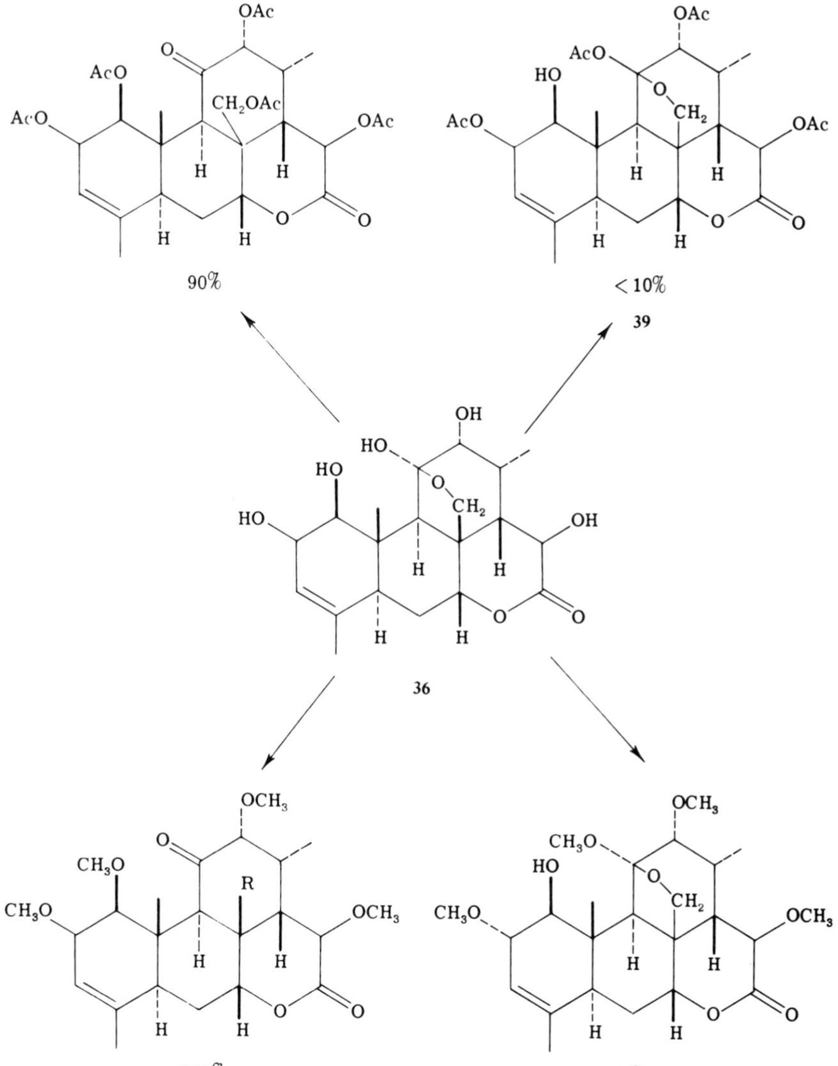

90%

< 10%

39

36

< 10%

R = CH_2OCH_3

38

90%

Methylation of these compounds according to Kuhn's method leads essentially to derivatives in which the hydroxyl of the hemiketal as well as all the hydroxyls present, except the one which is β-equatorial at C-1, have been methylated. Because of the steric hindrance caused by the presence of a ketal function at C-11, the hydroxyl at C-1 cannot be acetylated either.

Unlike methylation, acetylation gives principally derivatives that possess a ketone function at C-11 and a primary acetoxy group at C-8. Valuable information about the nature of the function present at C-11 (ketone or ketal) is provided by circular dichroism measurements as well as by the NMR data of the proton H_9 and those of the -CH_2O- grouping at C-8.

In the case of glaucarubol (**36**), we were able to isolate a small amount of a methylated derivative (**38**) with a keto grouping, as well as a small quantity of an acetate (**39**) with a hemiketal. But it should be noted that

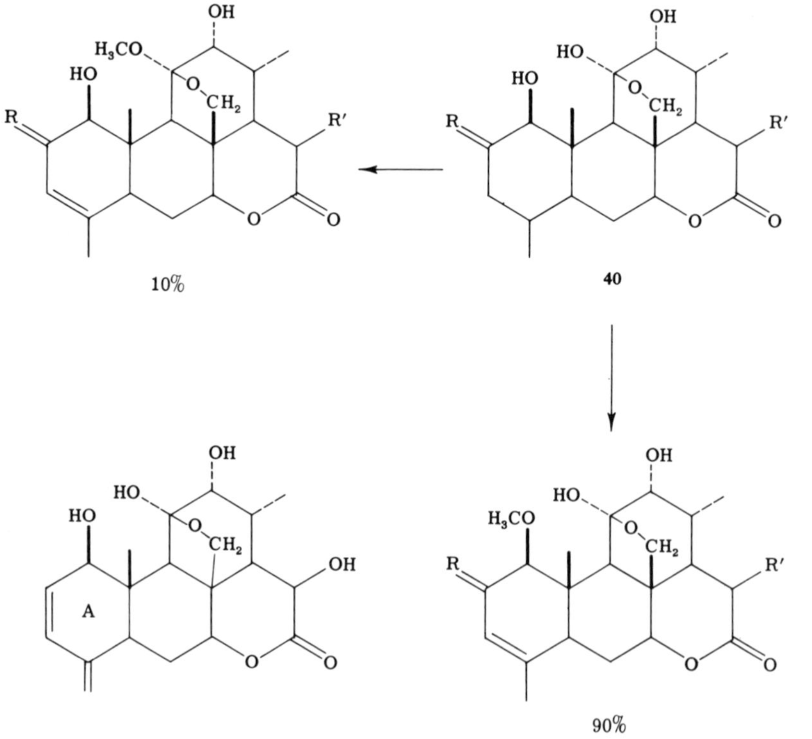

10%

40

90%

a: R = O, ; R′ = H
b: R = H, OH; R′ = OH
c: R = H, OH; R′ = H
d: R = O, ; R′ = OH

only the derivatives that have a carbonyl at C-11 are fully acetylated or methylated, that is to say, *penta*-acetylated or *penta*-methylated. On the contrary, when the ketal is present, the derivatives are *tetra*-methylated or *tetra*-acetylated. In other words, when the ketal is present at C-11, the β-equatorial hydroxyl at C-1 is sterically hindered so that it does not react upon acetylation or methylation.

3. There is in the chemistry of this type of compound an unexpected functional group interaction which merits mention. This is the reaction with diazomethane (Gaudemer *et al.*, 1967). All the quassinoids with structure **40** give, on treatment in methanol with ethereal diazomethane, two monomethylated derivatives in the ratio of about 9 to 1. In the minor product the hydroxyl of the hemiketal has been methylated whereas the hydroxyl at C-1 has been methylated in the major product. It has been shown that this reaction is independent of the nature of the substituent at C-2 or C-12. The formation of the products, particularly of that with the methoxyl at C-1, may be explained by strong hydrogen bonding between the hydroxyls at C-1 and C-11; this view is supported by the high value of the $\Delta\nu$(OH) observed in the infrared spectra of some of these compounds.

NMR spectroscopy was without doubt the most important physical tool in the structural elucidation of the quassinoids. The presence of numerous and varied functional groups in these molecules causes the protons to be shielded to different extents. Thus, many of the protons appear in distinctive regions of the spectrum leading to quite characteristic NMR patterns for this kind of compound.

Thus, the —CH_2O—grouping at C-8 generally gives rise to a two-proton nonequivalence quartet (AB systems); the chemical shift of these protons, as well as their coupling constant, are dependent on the nature of the function in which they are involved (hemiketal, ketal, acetoxy, or oxide). The β-equatorial $H_{(7)}$ proton appears in the spectra of the quassinoids which have a δ-lactone as a characteristic and easily recognizable triplet ($J \approx 5$ cps). In the compounds having a hydroxyl or esterified hydroxyl at position 15, the α-axial proton $H_{(15)}$ ($J_{15,14} \approx 10$ cps) is found downfield from the range normally accepted for such protons; this is probably due to the closely located oxygen function at C-12. Moreover, the chemical shift of proton $H_{(15)}$ depends upon the nature of the group at C-11: the $H_{(15)}$-resonance is shifted downfield when a hemiketal or a hydroxyl at C-11 is replaced by a keto group. $H_{(15)}$ appears as a doublet with a large coupling constant of about 10 cps since it is diaxially coupled with proton $H_{(14)}$.

In contrast to NMR spectroscopy, little use was made of mass spectroscopy in the early structural studies of the quassinoids. But the method became more applicable with the introduction of improved inlet systems that permit the vaprorization of relatively nonvolatile substances, such as the quassinoids. Yet, because of the ignorance of the cracking pattern of this kind of molecule, mass spectroscopy was mainly used for molecular weight determinations. This was of great importance since elemental analysis was not reliable for establishing molecular formulas because of ready solvation during crystallization. Recently, a systematic study of the mass spectra of several of these compounds was made (Fourrey *et al.*, 1968) and fragmentation patterns characteristic for some of their structural features were observed.

Most of the mass spectral studies have been made on ailanthone (**20**) and its numerous derivatives, particularly the 1-*O*- and 11-*O*-alkyl derivatives. The principal fragments in the mass spectrum of ailanthone are found at *m/e* 135, 151, and 248, corresponding to ions y′, $[C_9H_{11}O]^+$; y, $[C_9H_{11}O_2]^+$;

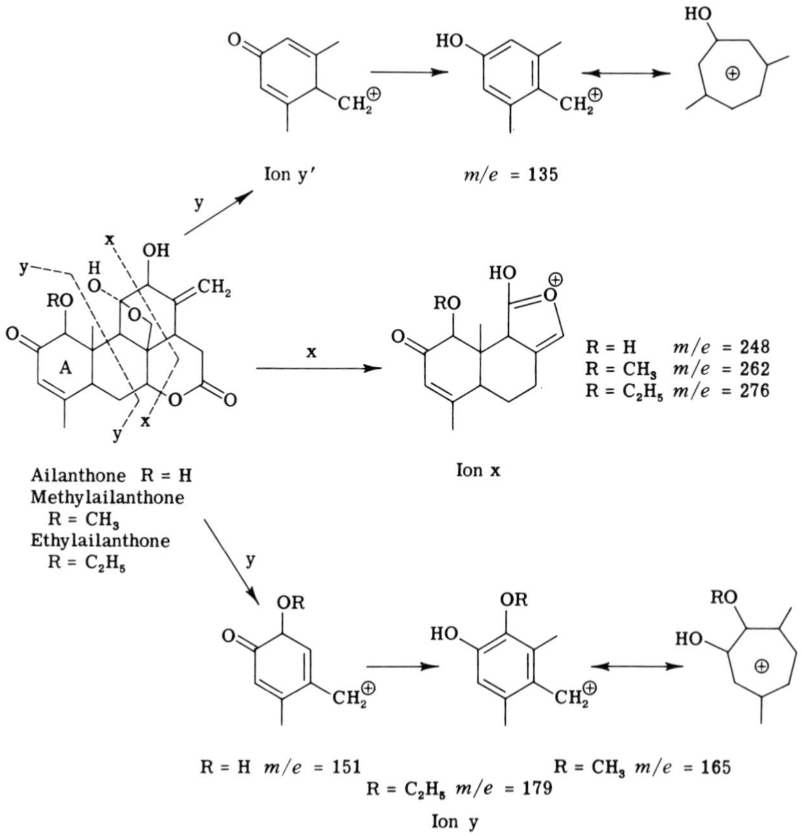

Ion y′ *m/e* = 135

Ailanthone R = H
Methylailanthone R = CH_3
Ethylailanthone R = C_2H_5

R = H *m/e* = 248
R = CH_3 *m/e* = 262
R = C_2H_5 *m/e* = 276

Ion x

R = H *m/e* = 151
R = CH_3 *m/e* = 165
R = C_2H_5 *m/e* = 179

Ion y

FIG. 1

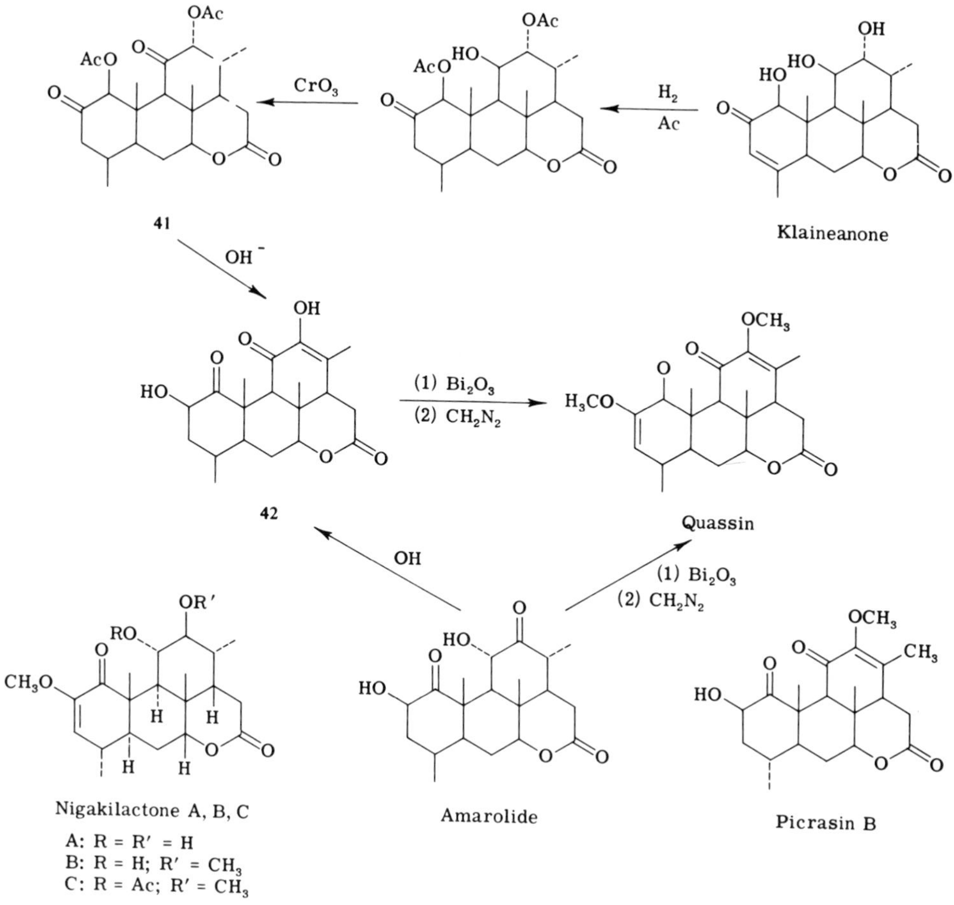

OAc
AcO
CrO_3
HO
H_2
Ac
OH
41
Klaineanone
OH^-
OCH_3
HO
(1) Bi_2O_3
(2) CH_2N_2
H_3CO
42
Quassin
OH
OR′
RO
CH_3O
H
Nigakilactone A, B, C
A: R = R′ = H
B: R = H; R′ = CH_3
C: R = Ac; R′ = CH_3
Amarolide
CH_3
Picrasin B

FIG. 2

and x, $[C_{14}H_{18}O_4]^+$, respectively (Fig. 1). In the mass spectra of 1-*O*-methylailanthone and 1-*O*-ethylailanthone, the positions of the latter two peaks are shifted by 14 and 28 mass units, respectively, while the *m/e* peak at 135 remains.

While ions of the type x are independent of the A-ring structure, the oxidation level (or the substitution pattern) in ring A seems to play a major role in the formation of ions y′ and y. This fragmentation is observed only when a hydroxyl is present at C-11. The results obtained by deuteration experiments indicate a transfer of a hydrogen of this hydroxyl to ring A.

Correlations of Some Quassinoids

None of these quassinoids has been synthesized so far. Some correlations have been carried out, and this by means of very simple reactions.

Thus, Casinovi and co-workers (1965) have correlated amarolide (**21**) to quassin (**1**) by oxidation with Bi_2O_3 to the bisdiosphenol which has been methylated with diazomethane; the last step can be reversed with boron tribromide. On the other hand, we converted klaineanone to quassin, and therefore in theory to amarolide, by the route shown in Fig. 2 (Zylber, 1968). Alkaline treatment of compound **41**, which is easily obtained from klaineanone, leads to the monodiosphenol (**42**). Thus, in addition to the hydrolysis of the acetoxy groups, the basic treatment of compound **41** causes oxidation at C-12 and a α-ketol rearrangement in ring A. Amarolide or acetylamarolide, when treated with base, also give the monodiosphenol (**42**). Oxidation of the latter with Bi_2O_3 followed by methylation with diazomethane then leads to a mixture of quassin and isoquassin which can be separated by chromatography.

Correlations between chaparrin, chapparinone, ailanthone, and glaucarubin have also been made (Fig. 3) (Zylber, 1968). Chapparin was converted to chaparrinone with manganese dioxide. Chaparrinone and glaucarubin have been interrelated through a glaucanol derivative. Treatment of dihydrochaparrol with acetic anhydride leads to the cyclic enol ether, which on ozonolysis affords compound (**43**). This was found to be identical to the acetylation product of the aldehyde formate obtained by periodic acid oxidation of dihydroglaucanol.

Biosynthesis

The biogenesis of the bitter substances of the quassin group has been amply discussed during recent years.

Valenta *et al.* (1961) formulated two fundamentally different biogenetic hypotheses for the formation of quassin (**1**) and neoquassin (**2**), the first Simaroubaceae constituents to have their structures established, also by these authors. The compounds could arise either from the diterpenoid pimaran skeleton (A) by a series of C-methyl shifts and a shift of a two-carbon fragment, or by an oxidative coupling of two identical C_{10} units (B); the distribution of the oxygen atoms and the obvious structural symmetry of these compounds seemed to favor the latter hypothesis.

As we have seen, there has been since 1961 a rapid advance in the structural studies of the simaroubaceous bitter constituents bringing to light a group of compounds with a close relationship to the quassin structure.

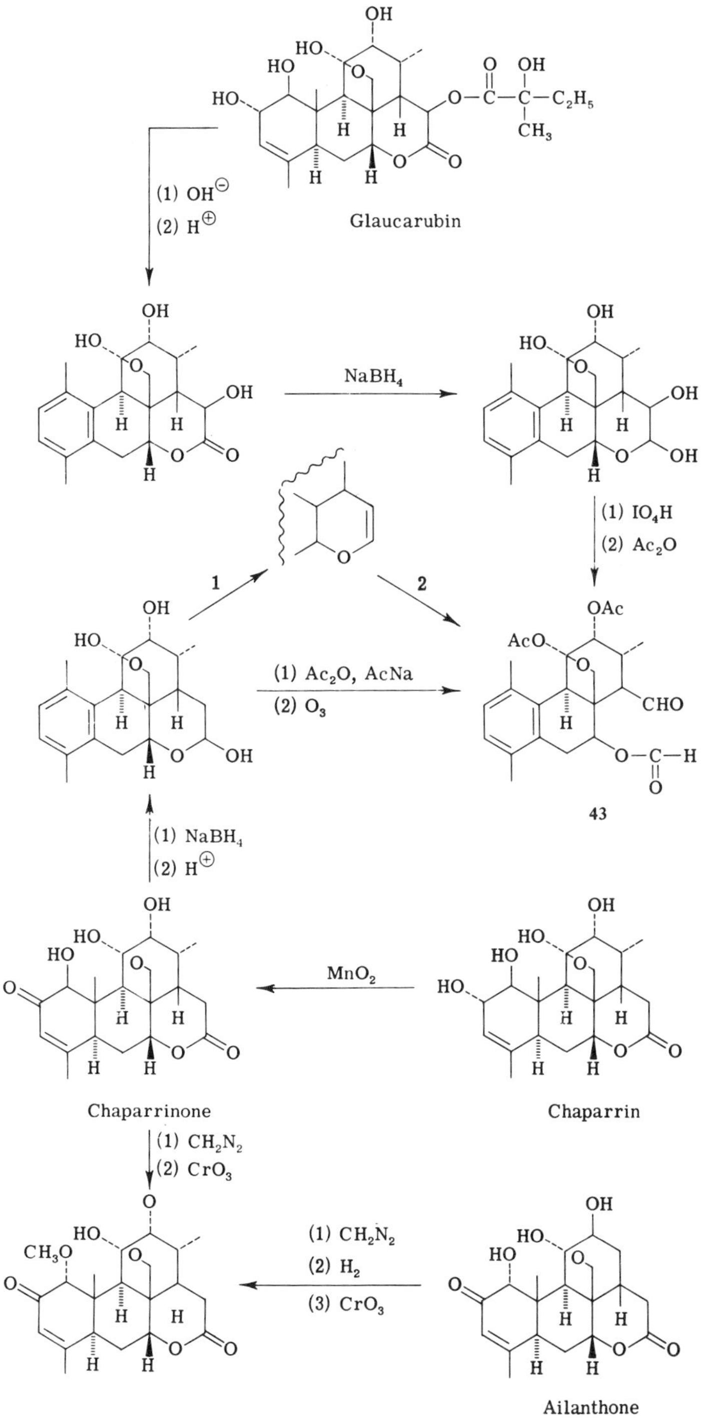

Glaucarubin

(1) $OH^{\ominus}$
(2) $H^{\oplus}$

$NaBH_4$

(1) IO_4H
(2) Ac_2O

1

2

(1) Ac_2O, AcNa
(2) O_3

43

(1) $NaBH_4$
(2) $H^{\oplus}$

MnO_2

Chaparrinone

Chaparrin

(1) CH_2N_2
(2) CrO_3

(1) CH_2N_2
(2) H_2
(3) CrO_3

Ailanthone

Fig. 3

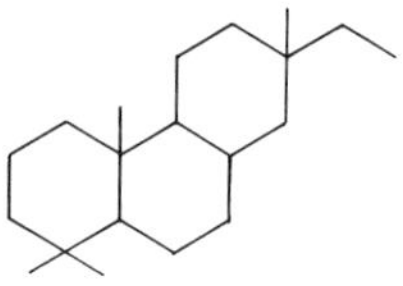

(A) 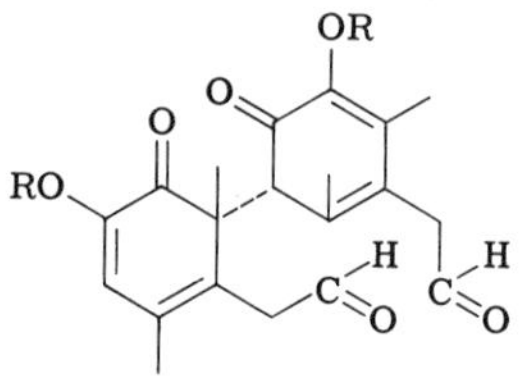 (B) 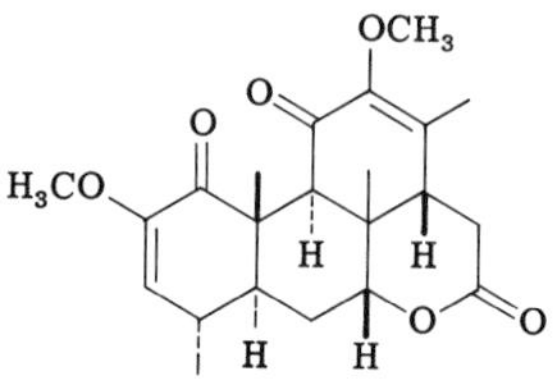 1

Their structural similarity indicates that there may be a single biogenetic pathway for all of them. A third biosynthetic pathway for the quassinoids has since been suggested by several authors (Polonsky *et al.*, 1964; Halsall and Aplin, 1964; son Bredenberg, 1964; Dreyer, 1964; Moss, 1966). According to this hypothesis, the quassinoids might follow a triterpenoid pathway similar to that formulated for the limonoids (Arigoni *et al.*, 1960). Thus, the hypothetical triterpenoid precursor of the quassinoids might also be apoeuphol or the 20α-isomer apotirucallol (**44**). Although the side chain is completely missing in most of the Simaroubaceae constituents, the exceptions, the C_{25} compounds, simarolide (**10**) and picrasin A (**11**) with a γ-lactone, are reminiscent of the furan ring present in the limonoids. Moreover, simarolide (**10**), whose structure has been confirmed by X-ray studies, has the absolute configuration shown, which is that of the triterpenoids.

The formation of the quassinoids from the prototriterpene apotirucallol (**44**) may be schematized as follows (Polonsky, 1964): a methyl group at C-4 and four carbon atoms at the end of the side chain are removed and carbons C-20 to C-23 are then converted into a γ-lactone ring. As supposed for the limonoids, ring D is oxidatively cleaved, via the α,β-unsaturated ketone (**45**), by a Baeyer-Villiger type oxidation. Opening of the δ-lactone (**46**), relactonization to the 7α-hydroxyl group followed by oxidation of the 17-hydroxyl would then lead to the basic skeleton of simarolide (**10**).

The C_{20} and C_{19} quassinoids would be formed by cleavage of the C-13–C-17 bond, the C_{19} compounds requiring furthermore the loss of carbon atom C-16. Simarolide is the only simaroubaceae bitter constituent which has no oxygenated function at C-12. This seems to indicate that in the other quassinoids the C-13–C-17 bond has been broken by a β-dicarbonyl system (12,17-dione), or by a retroaldol reaction of a 12-one-17-ol. But the cleavage of the C-13–C-17 by a Baeyer-Villiger oxidation of the 17-ketone should also be considered; the presence of an oxygenated group at C-13 in several quassinoids may be relevant in this context.

In order to verify the validity of one of the three hypotheses, we made

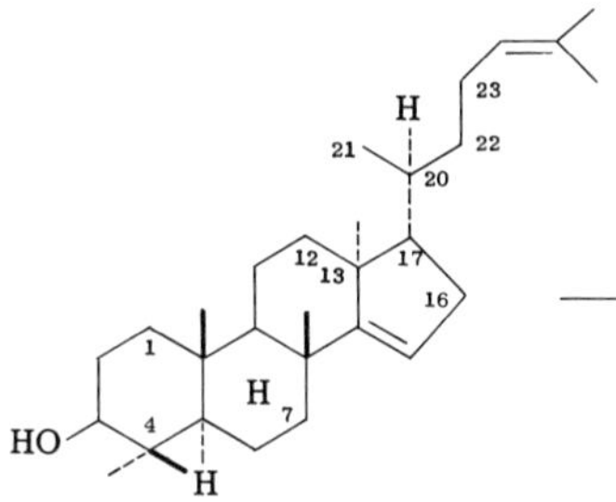

Apotirucallol

44

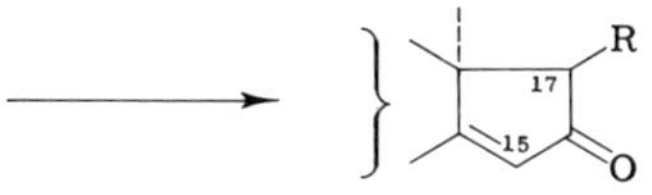

45

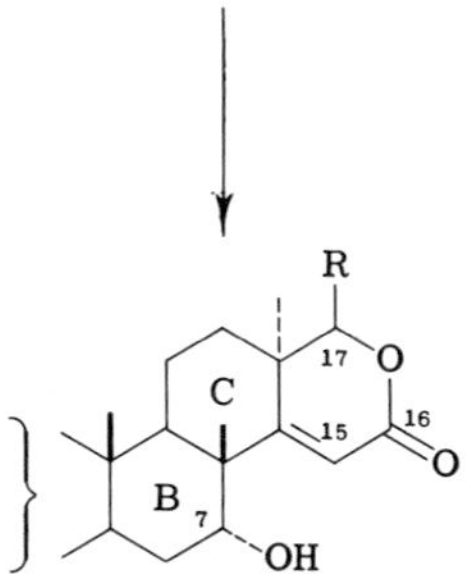

46

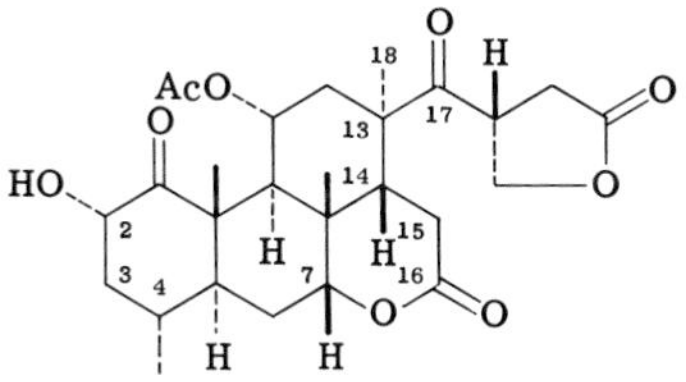

Simarolide

10

biogenetic experiments with labeled mevalonic acid lactone (**45**), one of the most active intermediates in the biogenesis of polyprenoids.

For our experiments we have used both 2-^{14}C-(*2)- and 5-^{14}C-(•5)-labeled mevalonic acid lactone (**45**) (Moron and Polonsky, 1968a).

Figure 4 shows the distribution of the radioactivity in a di- and triterpene biosynthesized from mevalonic lactone labeled at position 2 (designated by stars) and from mevalonic lactone labeled at position 5 (designated by dots). The difference concerns essentially the carbons 12 and 15.

Our biogenetic experiments have been carried out on glaucarubinone and glaucarubolone, the bitter constituents of *Simarouba glauca.* We obtained viable seeds of this member of the Simaroubaceae, not without difficulty, from San Salvador and from Guatemala.

Labeled mevalonolactone was incorporated by the young shoots (15 days old) of *Simarouba glauca,* placed in groups of 10 in small beakers. Eight days later, radioactive glaucarubinone and glaucarubolone were isolated.

In Fig. 5 we see the distribution of the radioactivity in glaucarubinone (**46**) and (**47**) when synthesized from a di- and triterpenoid precursor. With 2-^{14}C-(*2)-labeled mevalonic acid lactone, carbon 12 is labeled when originating from a diterpenoid whereas it is the carbon 15 which is labeled in the case of a triterpenoid precursor.

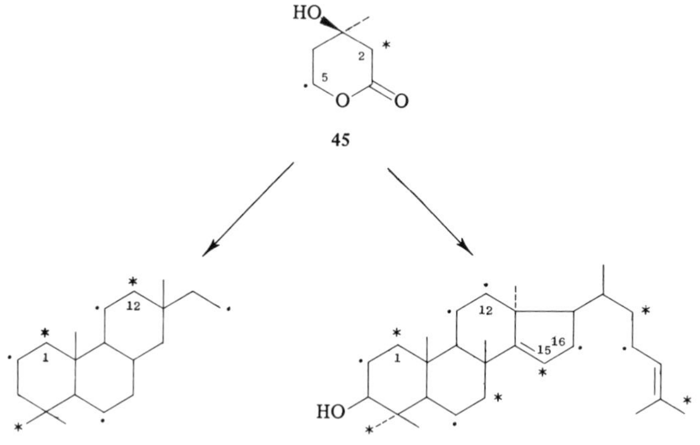

45

FIG. 4

With mevalonolactone-5-^{14}C, there must be *five* labeled carbons in those compounds originating from a triterpene whereas only four are labeled in the case of a diterpenoid origin. The additional carbon is the C-12 carbon atom.

Table 3 summarizes the distribution of radioactivity in glaucarubolone having a di- or triterpenoid origin biosynthesized from mevalonic acid labeled at positions 2 and 5. Localization of the radioactivity was achieved through degradation procedures carried out on this radioactive material.

For the required degradations (Fig. 6), the experience gained in the course of the structural work of these simaroubaceous constituents was of great help, although occasionally new procedures had to be devised. Hydrolysis of the radioactive glaucarubinone (**15**) isolated from both series of experiments showed that the α-hydroxy-α-methylbutyric acid residue was radioinactive and that the label was confined to the C_{20} component, glaucarubolone (**16**). It may be noted that isoleucine has proved to be the biogenetic precursor of the hydroxy acid (Moron and Polonsky, 1968b).

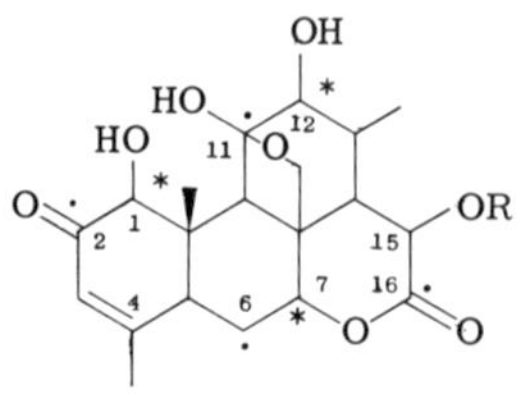

46 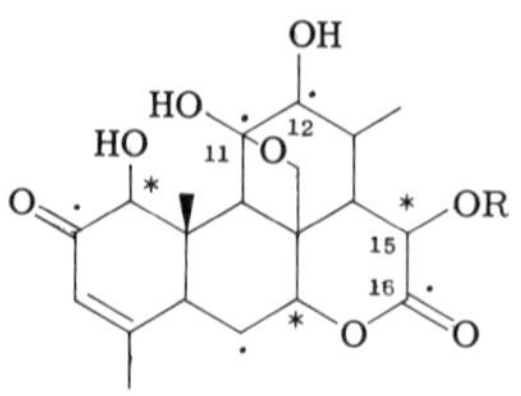 47

FIG. 5

TABLE 3

DISTRIBUTION OF RADIOACTIVITY IN GLAUCARUBOLONE BIOSYNTHESIZED FROM LABELED MEVALONIC ACID

	Precursor			
	[2-^{14}C]Mevalonic acid		[5-^{14}C]Mevalonic acid	
Carbon atom	Diterpene	Triterpene	Diterpene	Triterpene
C-1	+	+		
C-2			+	+
C-6			+	+
C-7	+	+		
C-11			+	+
C-12	+			+
C-15		+		
C-16			+	+
$C_{(4)}$-CH_3	+?	+?		

Before the oxidative degradation of the mevalonate-2-^{14}C-labeled glaucarubinone or glaucarubolone was studied, it was important to know whether the labeled or the unlabeled methyl of the *gem*-dimethyl group of the hypothetical terpenoid presursor (**44**) was retained at position 4 in glaucarubolone. Acetic acid obtained upon Kuhn-Roth oxidation of 2-^{14}C MVA-derived glaucarubolone proved to be radioinactive. The three methyl groups and the carbon atoms C-4, C-10, and C-13 are therefore unlabeled. The methyl group at position 4 in glaucarubolone must then be derived from the methyl group of mevalonate and must correspond to the β axial methyl group at C-4 of the terpenoid precursor (**44**) (Arigoni, 1959), since it is known that the α-methyl group corresponds to the carbon 2 of mevalonic acid.

The label of mevalonolactone-2-^{14}C was specifically incorporated into the carbon atom C-1, since Kuhn-Roth acetic acid from glaucarubolone was inactive, whereas the acetic acid from its partially aromatized derivative, dihydroglaucanol (**47**), was radioactive. This was further confirmed by the tetraesters (**48** and **49**) obtained by nitric acid oxidation of dihydroglaucanol. When obtained from dihydroglaucanol-2-^{14}C, each ester carried one third of the original label; when obtained from dihydroglaucanol-5-^{14}C each ester had two-fifths of the total radioactivity.

C-12 was isolated as follows: methylation of glaucarubinone and oxidation gave the acid (**50**). Decarboxylation afforded CO_2 unlabeled when

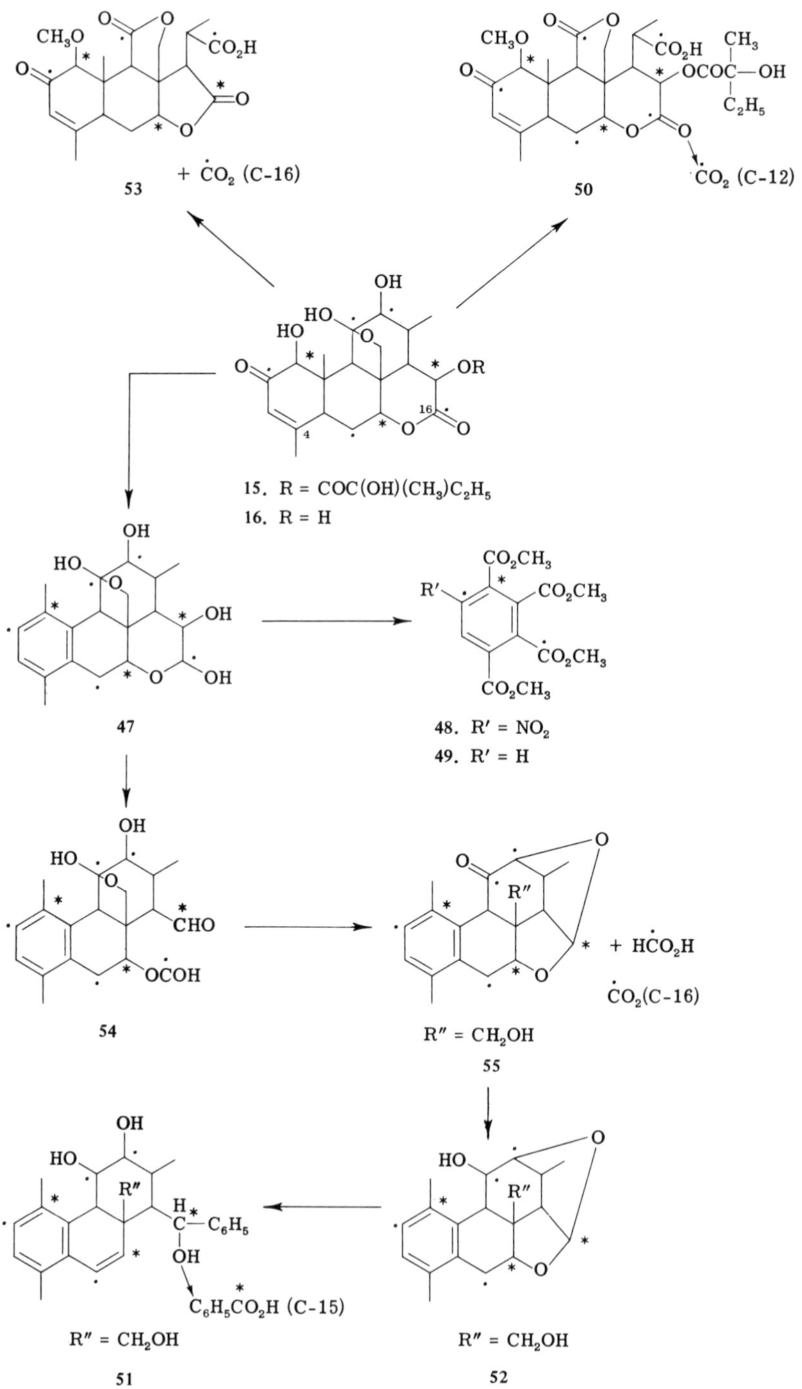

53

+ $\dot{C}O_2$ (C-16)

50

$\dot{C}O_2$ (C-12)

15. R = $COC(OH)(CH_3)C_2H_5$

16. R = H

47

48. R′ = NO_2

49. R′ = H

54

R″ = CH_2OH

55

+ $H\dot{C}O_2H$

$\dot{C}O_2$(C-16)

$C_6H_5\overset{*}{C}O_2H$ (C-15)

R″ = CH_2OH

51

R″ = CH_2OH

52

Fig. 6

derived from (*2)-glaucarubinone, but carrying one-fifth the original label present in (•5)-glaucarubinone.

C-15 was isolated as benzoic acid by Kuhn-Roth oxidation of the phenyl carbinol (**51**) obtained from the ketal (**52**) by phenyl lithium. The benzoic acid carried, as required, one-third the original label when derived from (*2)-glaucarubinone.

C-16 was isolated either by oxidation of 1-*O*-methylglaucarubolone to the 16-nor acid (**53**) and CO_2 or by acidic hydrolysis of the formate (**54**) to the ketal (**55**) and formic acid. C-16 thus obtained was found to be labeled from (•5)-but unlabeled from (*2)-glaucarubolone.

The labeling patterns observed in 2-^{14}C MV- and 5-^{14}C MV-labeled glaucarubinone as well as the presence of five labeled carbon atoms in the latter prove unambiguously the triterpene origin of these compounds. Actually, they exclude not only the diterpernoid origin of these quassinoids, but also the first hypothesis, that is to say, the oxidative coupling of two C-10 phenolic units which may originate from a monoterpene, and this regardless of the arrangement of these two units.

The results obtained are consistent with the scheme involving a tetracyclic triterpene, such apoeuphol or apotirucallol as a precursor of these simaroubaceous bitter constituents. Thus, the quassinoids are to be regarded not as diterpenoids, but as degraded triterpenoids.

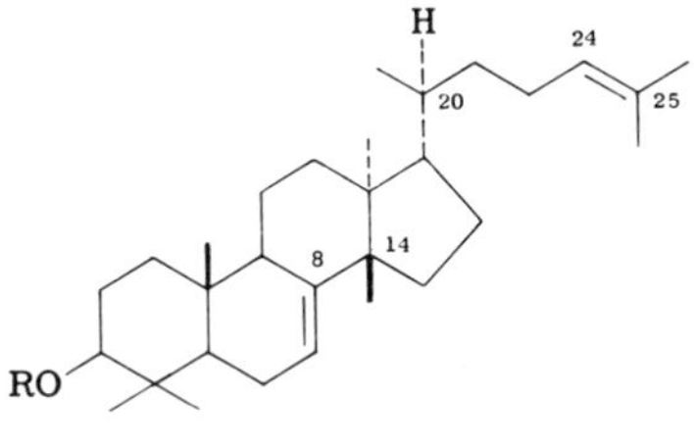

57

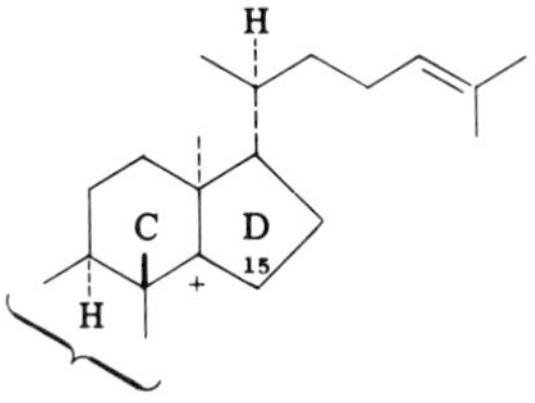

56

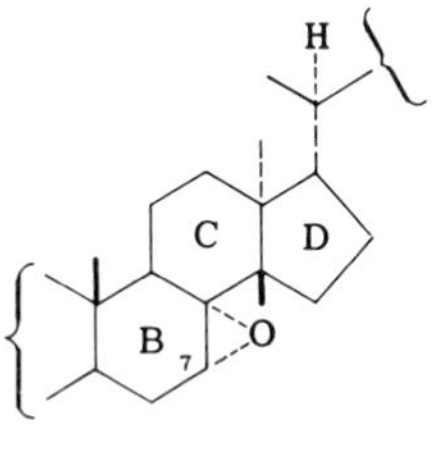

58

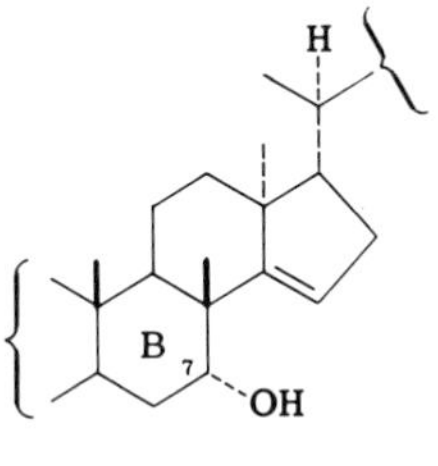

59

As to the prototriterpene, apotirucallol (**44**), or the 20-epimer, apoeuphol, it might be biosynthesized either directly from epoxidosqualene via the ion (**56**) and subsequent loss of a proton from C-15 or via Δ^7-tirucallol (**57**) which undergoes a skeletal rearrangement during which one methyl group migrates back from C-14 to C-8. This last step could be triggered by oxidative attack of the nuclear double bond of the Δ^7-isomer of tirucallol or euphol.

Evidence that such rearrangements of the C-14 methyl group can be brought about is provided by the rearrangement of 7,8-α-epoxidotirucallol derivatives (**58**) to 7-α-hydroxy compounds in the apotirucallol series (**59**) (Buchanan and Halsall, 1969).

The suggestion that a derivative of tirucalla-7,24-dien-3-ol (**57**) might be a precursor of the quassinoids (and limonoids) is supported by the occurrence of a number of side chain-oxygenated tirucall-7-ene derivatives in species of families from which limonoids and meliacins have been isolated. But it is only quite recently that two of these triterpenes have been isolated from a simaroubaceous species. They are melianodiol and its diacetate, isolated from *Samadera madagascariensis* (Merrien and Polonsky, 1971). Melianodiol (**60**) was previously isolated from *Melia azadirachta* (Meliaceae) and has been obtained by acid-catalyzed opening of the epoxide of melianone (**61**) by Lavie *et al.* (1967), The isolation of the diacetate indicates that melianodiol is not an artefact formed from melianone during the extraction.

We carried out, on melianodiol, some biogenetic-type reactions directed toward simarolide. Thus, the side chain of melianodiol (**60**) has been converted into a γ-lactone (**64**), and this by the following reactions: treatment of melianodiol with sodium borohydride yields the pentaol (**62**), which is cleaved by metaperiodate to the acetal (**63**). Oxidation with silver carbonate (Fétizon and Golfier, 1968) leads then to compound **64** with a γ-lactone such as is found in simarolide (**10**). On acetylation it gives the acetate (**65**).

As mentioned above, the quassinoids as well as the limonoids, which are all oxygenated at C-7, could arise, according to one of the two biogenetic hypotheses, by the formation of the 7α,8α-epoxide of a tirucall-7-ene derivative followed by rearrangement to a 7α-hydroxy-Δ^{14}-apo derivative. Such transformations have now been carried out on this lactone following the procedure of Buchanan and Halsall (1969). When it was treated with monoperphthalic acid in dry ether at 0°C for 48 hours, it yielded the 7α-,8α-epoxide (**66**) which rearranged with boron trifluoride etherate to give the apo compound (**67**).

It may be noted that treatment of compound **65** with *m*-chloroperbenzoic acid leads to the 7,8;9,11-diepoxide, whereas with *p*-nitroperbenzoic acid

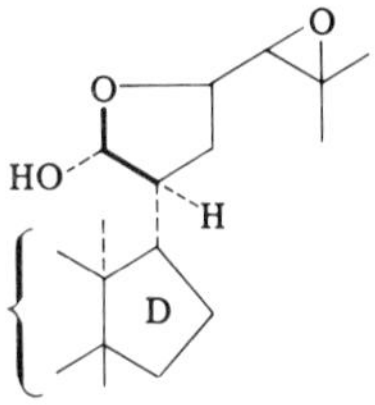

61

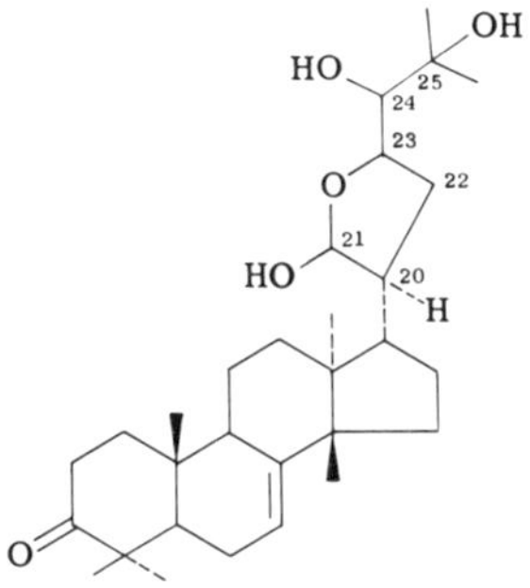

Melianodiol

60

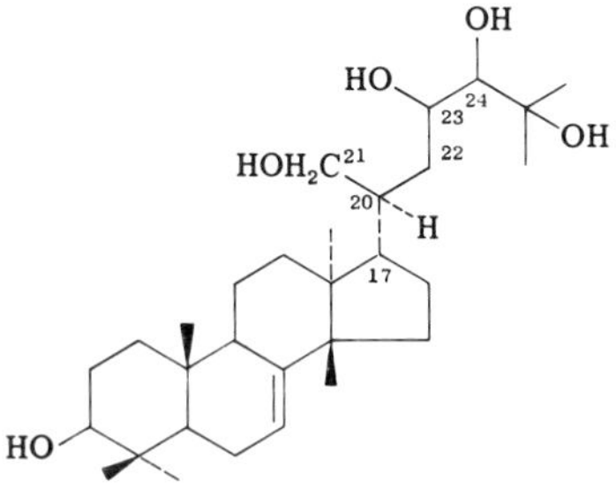

62

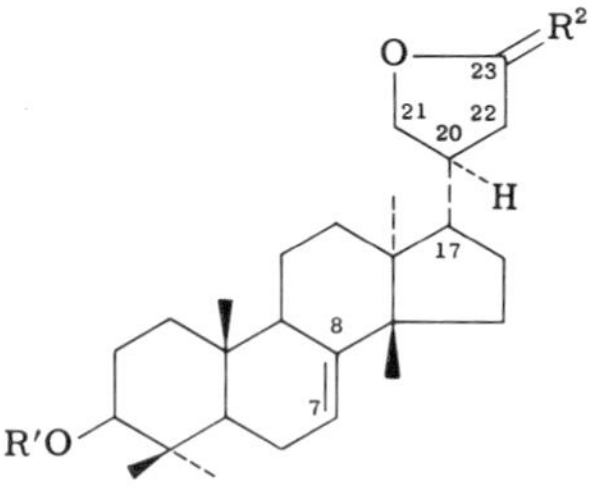

63. R′ = H; R^2 = OH; H
64. R′ = H; R^2 = O
65. R′ = Ac; R^2 = O

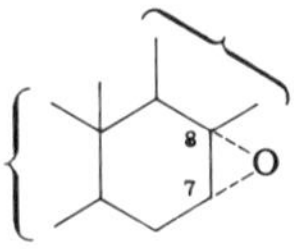

66

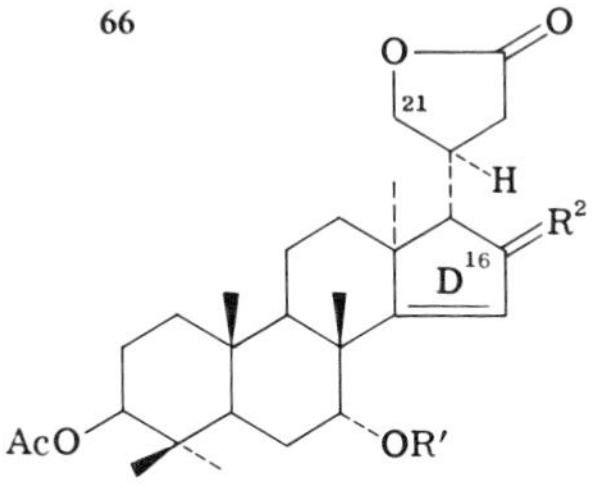

67. R′ = H; $R^2 = H_2$
68. R′ = Ac; $R^2 = H_2$
69. R′ = Ac; R^2 = O

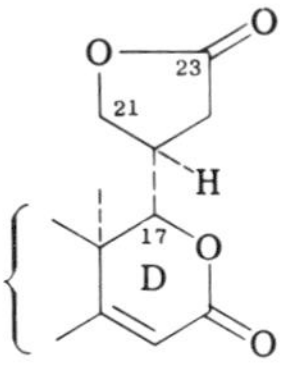

70

a mixture of the epoxide and the rearranged apo compound (**67**) was obtained.

Upon oxidation at 0°C with Jones' reagent, the latter gives a ketone and upon acetylation it affords the diacetate (**68**), which, upon treatment with chromic acid in acetone, smoothly oxidizes to the α,β-unsaturated ketone (**69**). When submitted to a Baeyer-Villiger oxidation, the latter yields the α,β-unsaturated lactone (**70**).

The double bond of this lactone grouping could not be hydrogenated, so that we could not go further; that is to say, we could not relactonize it to the hydroxyl at position 7 (after opening the lactone).

The isolation of side chain-oxygenated tirucall-7-ene derivatives from a simaroubaceous plant as well as the transformations we report are clearly relevant to the biogenetic proposals that have been made for the quassinoids So it seems very likely that a Δ^7-tirucallene derivative is the prototriterpene of the quassinoids.

In order to get some insight into the mode of formation of the triterpenic precursor of the quassinoids, mevalonic acid labeled at position 4*R* with tritium (**71**) was incorporated into glaucarubinone and glaucarubolone by viable seeds of *Simarouba glauca* (Moron *et al.*, 1971).

We have seen that these quassinoids biosynthesized from mevalonic acid labeled with ^{14}C at position 5, possess five labeled carbon atoms (C-2, C-6, C-11, C-12, and C-16). When biosynthesized from mevalonic acid tritiated at position 4*R* (**71**), they should have, if all the tritium atoms were retained, three tritiums (see the encircled hydrogens), at position 3, 5, and 9. The atomic ratio, T:^{14}C, in these compounds when biosynthesized from both of these mevalonic acids should be 3:5.

The value of the atomic ratio of radioactive glaucarubinone and glaucarubolone, isolated from this experiment, indicates the presence of only two tritiums. They have been localized at position 5 and 9 as follows: dihydroglaucanol (**47**), obtained as already mentioned by reduction with sodium borohydride followed by aromatization, reveals a decrease in the atomic ratio T:^{14}C, in agreement with the elimination of proton 5. Acetylation of dihydroglaucanol under mild conditions yeilds the tetraacetate (**72**), which has the same atomic ratio, T:^{14}C, as dihydroglaucanol. Under more drastic conditions, dihydroglaucanol gives upon acetylation the enol acetate (**73**). Its atomic ratio shows that the elimination of proton $H_{(9)}$ is accompanied by the nearly total loss of tritium in the molecule. These results show that two tritium atoms—at positions 5 and 9—are retained in glaucarubolone when biosynthesized from mevalonic acid tritiated at position 4*R*. The absence of tritium at position 3 supports the hypothesis that the loss of the α-equatorial methyl at C-4 proceeds via a compound having a carbonyl at position 3, for instance by decarboxylation of a β-keto acid. The retention of tritium at position 9 excludes the involvement of a triterpene precursor with a $\Delta^{8,(9)}$ double bond and makes that of a 9,10-cyclopropane precursor unlikely. So, if $\Delta^{7,8}$-euphol or $\Delta^{7,8}$-tirucallol is the precursor of the quassinoids, they should be formed by stabilization of the intermediate carbocation at C-8 without the intervention of euphol or tirucallol.

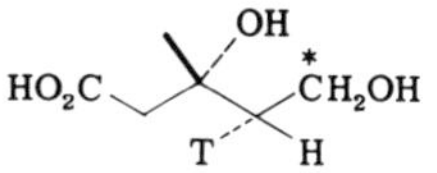

71

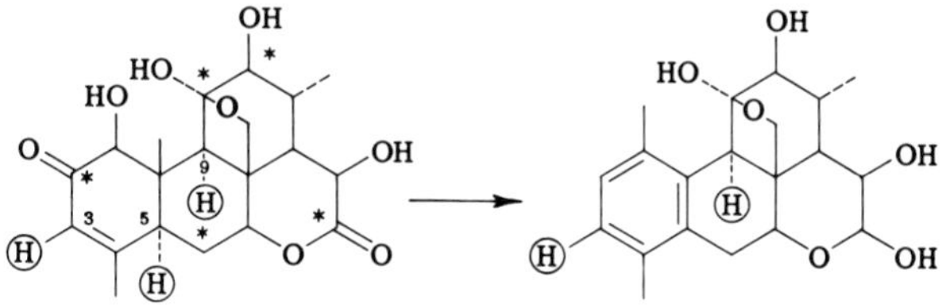

Glaucarubolone
(4*R*-4*T*, 5-^{14}C AMV)

16

47

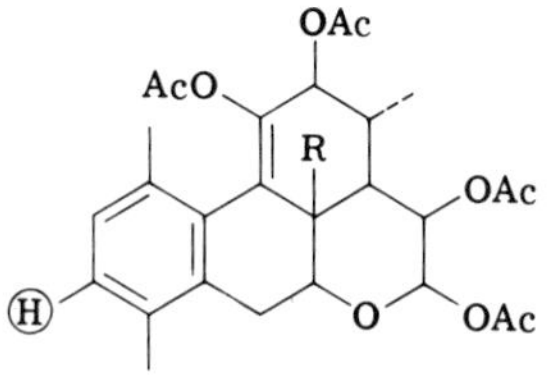

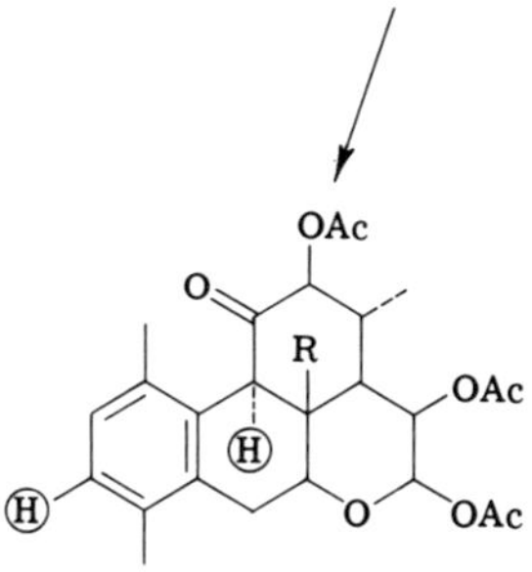

R = CH_2OAc

73

R = CH_2OAc

72

References

Arigoni, D. 1959. *Terpene Sterol Biosyn., Ciba Found. Symp.* p. 244.

Arigoni, D., D. H. R. Barton, E. J. Corey, O. Jeger, L. Caglioti, S. Dev, P. G. Ferrini, E. R. Glazier, A. Melera, S. K. Pradhan, K. Schaffner, S. Sternhill, J. F. Templeton, and S. Tobinga. 1960. *Experientia* **16**:41.

Beer, R. J. S., B. G. Dutton, D. B. G. Jaquiss, A. Robertson, and W. E. Savige. 1956. *J. Chem. Soc.* p. 4850.

Bredenberg, J. B.-son. 1964. *Chem. Ind.* (*London*) p. 73.

Brown, W. A., and G. A. Sim. 1964. *Proc. Chem. Soc.* p. 293.

Buchanan, G. St. C., and T. G. Halsall. 1969. *Chem. Commun.* p. 242.

Casinovi, C. G., P. Ceccherelli, G. Grandiloni, and V. Bellavita. 1964. *Tetrahedron Lett.* p. 3991.

Casinovi, C. G., V. Bellavita, G. Grandolini, and P. Ceccherelli. 1965. *Tetrahedron Lett.* p. 2273.

Clark, E. P. 1937. *J. Amer. Chem. Soc.* **59**:927, 2511.

Connoly, J. D., K. Overton, and J. Polonsky. 1970. *In* "Progress in Phytochemistry" (L. Reinhold and Y. Liwschitz, eds.), Vol II, pp. 385–455. Wiley (Interscience), New York.

Dreyer, D. L. 1964. *Experienta* **20**:297.

Duncan, G. R., and D. B. Henderson. 1968. *Experientia* **24**:768.

Fétizon, M., and M. Golfier. 1968. *C. R. Acad. Sci., Ser. C* **267**:900.
Fourrey, J.-L. 1968. These de doctorat, Faculté des Sciences d'Orsay de l'Université de Paris-Sud.
Fourrey, J.-L, B. C. Das, and J. Polonsky. 1968. *Org. Mass Spectrom.* **1**:819.
Gaudemer, A., and J. Polonsky. 1965. *Phytochemistry* **4**:149.
Gaudemer, A., J.-L. Fourrey, and J. Polonsky. 1967. *Bull. Soc. Chim. Fr.* p. 1676.
Geissman, T. A., and G. A. Ellestad. 1962. *Tetrahedron Lett.* p. 1083.
Halsall, T. G., and R. T. Aplin. 1964. *In* "Progress in the Chemistry of Organic Natural Products" (L. Zechmeister, ed.), p. 167. Springer-Verlag, Berlin and New York.
Hikino, H., T. Ohta, and T. Takemoto. 1970a. *Chem. Pharm. Bull.* **18**:219.
Hikino, H., T. Ohta, and T. Takemoto. 1970b. *Chem. Pharm. Bull.* **18**:1082.
Hollands, T. R., P. de Mayo, M. Nisbet, and P. Crabbé. 1965. *Can. J. Chem.* **43**:2996.
Lavie, D., M. K. Jain, and S. R. Shpan-Gabrielith. 1967. *Chem. Commun.* p. 910.
Lê-Van-Thoi and Nguyên-Ngoc-Suong. 1970. *J. Org. Chem.* **35**:1104.
Mattox, V. R. 1952. *J. Amer. Chem. Soc.* **74**:4340.
Merrien, M.-A., and J. Polonsky. 1971. *Chem. Commun.* p. 261.
Mitchell, R. E., W. Stöcklin, M. Stefanović, and T. A. Geissman. 1971. *Phytochemistry* **10**:411.
Moron, J., and J. Polonsky. 1968a. *Tetrahedron Lett.* p. 385.
Moron, J., and J. Polonsky. 1968b. *Eur. J. Biochem.* **3**:488.
Moron, J., M.-A. Merrien, and J. Polonsky. 1971. *Phytochemistry* **10**:585.
Moss, G. P. 1966. *Planta Med.* **14**:Suppl., 86.
Murae, T., T. Ikeda, T. Tsuyuki, T. Nishihama, and T. Takahashi. 1970a. *Bull. Chem. Soc. Jap.* **43**:969.
Murae, T., T. Ikeda, T. Tsuyuki, T. Nishihama, and T. Takahashi. 1970b. *Bull. Chem. Soc. Jap.* **43**:3021.
Murae, T., T. Ikeda, T. Tsuyuki, T. Nishihama, and T. Takahashi. 1970c. *Chem. Pharm. Bull. Jap.* **18**:2590.
Murae, T., T. Tsuyuki, T. Ikeda, T. Nishihama, S. Masuda, and T. Takahashi. 1971. *Tetrahedron* **27**:1545.
Polonsky, J. 1964. *Proc. Chem. Soc.* p. 292.
Polonsky, J., and N. Bourguignon-Zylber. 1965. *Bull. Soc. Chim. Fr.* p. 2793.
Polonsky, J., and J.-L. Fourrey. 1964. *Tetrahedron Lett.* p. 3983.
Polonsky, J., C. Fouquey, and A. Gaudemer. 1964. *Bull Soc. Chim. Fr.* p. 1818, 1827.
Polonsky, J., Z. Baskevitch, A. Gaudemer, and B. C. Das. 1967. *Experientia* **23**:424.
Polonsky, J., Z. Baskevitch, B. C. Das, and J. Müller. 1968. *C. R. Acad. Sci., Ser. C.* **267**:1346.
Polonsky, J., Z. Baskevitch, and J. Müller. 1969. *C. R. Acad. Sci. Ser. C.* **268**:1392.
Sim, K. Y., J. Sims, and T. A. Geissman. 1968. *J. Org. Chem.* **33**:429.
Stöcklin, W., and T. A. Geissman. 1968. *Tetrahedron Lett.* **57**:6007.
Stöcklin, W., and T. A. Geissman. 1970. *Phytochemistry* **9**:1887.
Stöcklin, W., L. B. de Silva, and T. A. Geissman. 1969. *Phytochemistry* **8**:1565.
Stöcklin, W., M. Stefanović, and T. A. Geissman. 1970. *Tetrahedron Lett.* **27**:2399.
Tresca, J. P., L. Alais, and J. Polonsky. 1971. *C. R. Acad. Sci., Ser. C.* **273**:601.
Valenta, Z., A. H. Gray, D. E. Orr, S. Papadopoulos, and C. Podešva. 1962. *Tetrahedron* **18**:1433.
Viala, B., and J. Polonsky. 1970. *C. R. Acad. Sci., Ser. C.* **271**:410.
Wall, M. E., and M. Wani. 1970. *In* Abstracts Book, 7th Int. Symp. Chem. Nat. Prod., IUPAC, Riga 614.
Zylber, N. 1968. Doctoral Thesis, Orsay, Paris.
Zylber, J., and J. Polansky. 1964. *Bull. Soc. Chim. France* p. 2016.
Bourguignon-Zylber, N., and J. Polansky. 1970. *Bull. Chim. Therap.* **6**:396.

THE BIOGENESIS OF SESQUITERPENE LACTONES OF THE COMPOSITAE

T. A. GEISSMAN

*Department of Chemistry, University of California at Los Angeles, Los Angeles, California**

Introduction

During the past few years a large number of new sesquiterpene lactones have been isolated in studies on plants closely allied botanically and in exhaustive examinations of the components of single plants. In view of the great many compounds of this class that are now known, questions of

* Contribution No. 2780 from the Department of Chemistry, University of California, Los Angeles.

structure alone have come to be of less interest than the meaning of structures in terms of their biosynthetic history and interrelationships.

Since it is a reasonable presumption that the 200-odd known lactones of this group have a common origin in synthesis in the plants—particularly those that co-occur or that are found in plants of close botanical affinity—it appears possible to derive from a critical examination of their structures some insight into the nature of the reactions involved in the transformation of simple compounds near the biosynthetic origin to complex compounds extensively modified in structure.

Classical methods of the study of biosynthesis remain to be applied to this group of compounds, for until adequate and versatile synthetic methods are devised for the preparation of suitably labeled candidate precursors, the direct demonstration of synthetic pathways will be deferred. This applies both to experiments with intact plants and with enzymes or enzyme preparations.

The present satisfactory state of understanding of many biosynthetic pathways has developed as the result of direct experiments with selected precursors and the isolation of the relevant products. Yet many, or most, such experiments are planned from considerations of the chemical plausibility and mechanistic feasibility of presumed pathways of synthesis. Before the advent of modern techniques, such predicted pathways provided the only clues to the understanding of the living processes, and, despite numerous mistaken predictions, a high degree of success attended such proposals, as shown by subsequent investigations using direct experimental methods.

The organic chemist, faced with a given structure, can easily derive a reaction sequence by which it might have been derived from a postulated precursor. The probability that the hypothesis represents the reality can be greatly increased if a series of compounds, all occurring together in a single organism, can be shown to be constituent parts of the hypothetical pathway. Indeed, the search for new compounds can often be guided by such conjecture.

The present paper represents an attempt to assess the probability of a number of biosynthetic transformations in the plants to be considered, and to propose some general reaction types that appear to be often involved in the biosynthetic processes that lead to the final products of synthesis. Such proposals as those to be offered, however compelling, are no more than the groundwork for further study and can have no more merit than that with which future findings may endow them.

The still unanswered questions of the biosynthesis of terpenoid compounds now lie largely at the level of secondary metabolic alterations, in the course of which simple precursors lying near the oxidation state of farnesol are modified by ring closures, rearrangements, and the introduction of oxygen functionality. There is general acceptance of the view that terpenoid compounds, from terpenes to polyterpenes and steroids, are the end products of a metabolic pathway that can be described in the following general terms:

acetate → mevalonate → isopentenyl pyrophosphates → geranyl pyrophosphate → farnesyl pyrophosphate → $(C_5)_x$ compounds.

Terpenes (C_{10}) arise from geraniol, sesquiterpenes (C_{15}) from farnesol, and so on. Several structurally and stereochemically consistent schemes of terpenoid biosynthesis by this route, supported (especially for triterpenes and steroids) by appropriate experiments in volving the use of labeled compounds, have been advanced, and there is now little doubt of the validity of these general concepts (Ruzicka, 1963; Hendrickson, 1959; Parker *et al.*, 1967).

Although there have been relatively few experimental demonstrations, with the use of labeled precursors, of the transformation of farnesyl pyrophosphate into sesquiterpenoid end products, such scanty evidence as does exist (Waldner *et al.*, 1969; Birch and Hussein, 1969; Anchel *et al.*, 1970; Staunton, 1969), coupled with the compelling nature of the hypothesis, permits the acceptance of the view that one important route of sesquiterpenoid biosynthesis can be represented as follows:

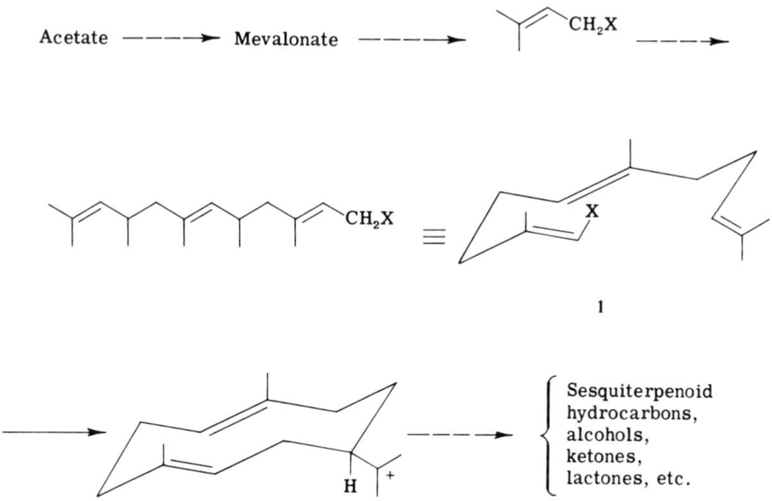

Acetate ----→ Mevalonate ----→ CH_2X ----→

CH_2X ≡ X

1

→ H + ----→ Sesquiterpenoid hydrocarbons, alcohols, ketones, lactones, etc.

2

The simplest of the monocyclic sesquiterpenes, hedycaryol (Jones and Sutherland, 1968), is in fact the tertiary alcohol derived from **2**:

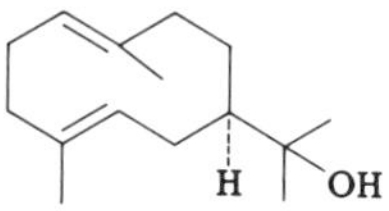

Hedycaryol

For these reasons, a consideration of much of sesquiterpenoid biosynthesis can reasonably begin by accepting the reality of the step **1** → **2**, and can be addressed to the processes by which, for example, **2** is converted into the variously cyclized and oxidized compounds represented, for example, by the variously constituted compounds costunolide (**3**), xanthinin (**4**), α-santonin (**5**), and ambrosin (**6**).

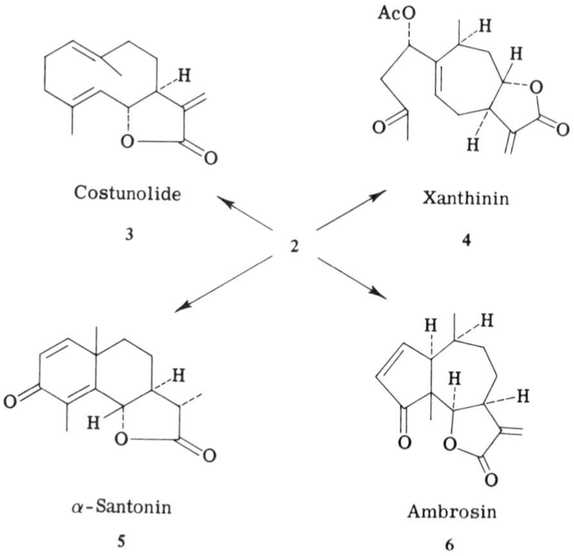

Costunolide **3** ← **2** → Xanthinin **4**

α-Santonin **5** ← **2** → Ambrosin **6**

The simplest of the sesquiterpene lactones in its relationship to farnesol is the germacranolide **3**, costunolide. Since costunolide is but a few steps removed from the primary precursor **2**, it is not surprising that it is found not only in several distinct sections of the family Compositae, but occurs in the quite unrelated Magnoliaceae as well. It is to be expected that compounds very near the primary levels of structural elaboration would be more widely distributed than those of more complex structure, for the latter are elaborated by successive steps which reflect the genetic individuality of the species in which they are found.

The C-11, C-12, and C-13 Atoms of the Lactone Ring

The formation of costunolide (**3**) from **2** requires two changes: (a) the introduction of oxygen into position 6, and (b) the oxidative transformation of the isopropenyl group (the deprotonated side chain of **2**) in the following way:

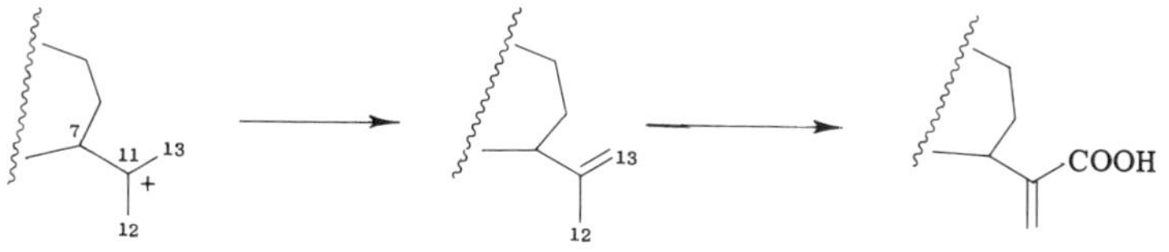

7 11 13 + 12 → 13 12 → COOH

2 7 8

What is the nature of the six-electron oxidation that leads from **7** to **8**? Is the —COOH group of **8** derived from the C-12 or C-13 atom of **7**? Although a decisive answer to this question cannot be arrived at by conjecture, it is suggestive to examine the various forms in which the lactone ring exists in nature. These can be represented as follows*:

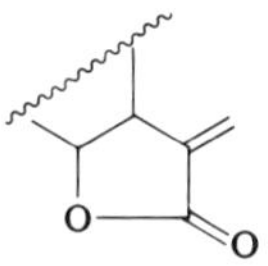

O O O O O O

9 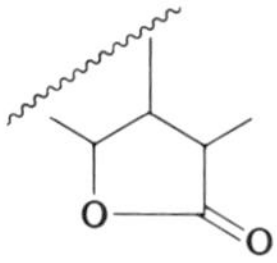 10 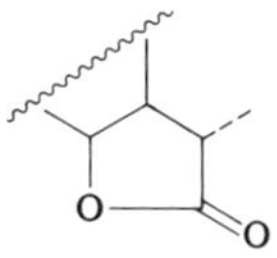 11

OH O O CH_2OH O O

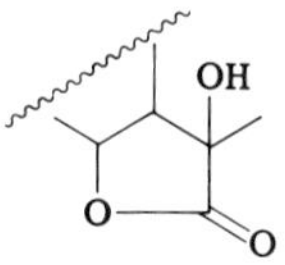

12 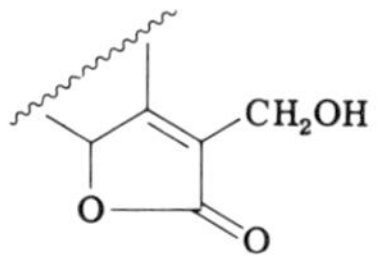 13

There are obviously a number of ways in which these structures can be derived by formal analogy to oxygenation, hydrogenation, and dehydrogenation processes that are known to occur in nature. Such formulations of unrelated processes would, however, have less appeal and carry less conviction than one in which a central primary process could serve as the focal point from which the formation of **9**, **10**, **11**, **12**, and **13** could ensue.

* Carboxylic acids corresponding to the opened form of lactone 9 are also known as natural products.

(a)

7 → 14 —H^+→ 15 (CH_2OH)

15 → 16 (CH_2OH) —(ox) (−2e)→ 17 (CHO) —(ox) (−2e)→ 18 (COOH)

(b)

14 —H^+ (rearr.)→ 19 (CHO) —(ox) (−2e)→ 20 (COOH)

(c)

16 (CH_2OH) → 21 (CH_2OH, O) —H^+→ 22 (CH_2OH, CH_2OH)

22 → 23 (CH_2OH, COOH)

As will be described in detail below, it is probable that an important primary natural process for the introduction of oxygen in compounds of many kinds is the epoxidation of a carbon-carbon double bond. It will be apparent that the possible transformations of an epoxide of **7** provide rational routes to all the structures **9–13**.

That the series **16** → **17** → **18** is a likely one is suggested by the co-occurrence of the compounds costol, costal, and costic acid in costus root (*Saussurea lappa*), although the formation of **16** through the steps **7** → **14** → **15** is not suggested by this observation in itself. It will be noted also that the furano sesquiterpenes found in *Ligularia* spp. and *Senecio* spp. can also derive from a precursor such as **22**.

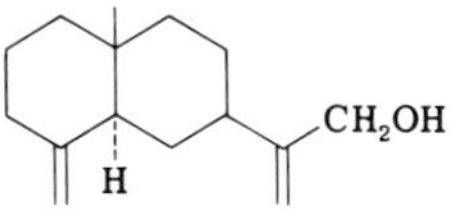

CH_2OH H — Costol

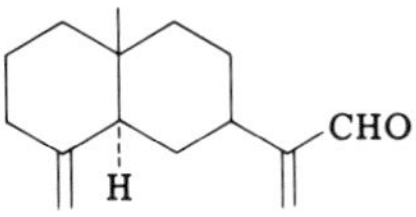

CHO H — Costal

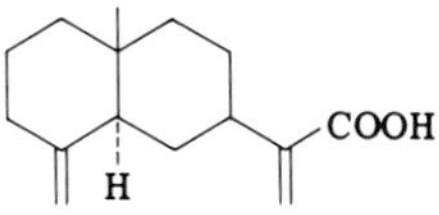

COOH H — Costic acid

It is implicit in the above suggestion that in the conversion **7** → **8** it is the carbon atom of $=CH_2$ (C-13) that becomes the carbonyl group of the lactones **9**, etc. This supposition remains to be tested by appropriate experiments with specifically labeled compounds.

The Introduction of Oxygen at C-6 and C-8

The question of the manner and time of introduction of the lactonic oxygen atom at C-6 and C-8 does not appear to be amenable to fruitful conjectures based upon straightforward mechanistic consideration. Compounds of the class are known with oxygen at C-6 only, at C-8 only, at both C-6 and C-8, all of them in both possible configurations. Some examples are the following:

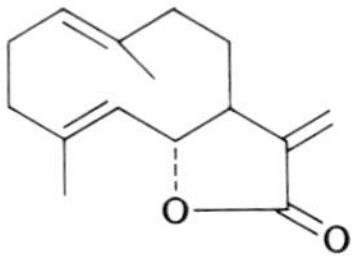

Costunolide
C-6-α
3

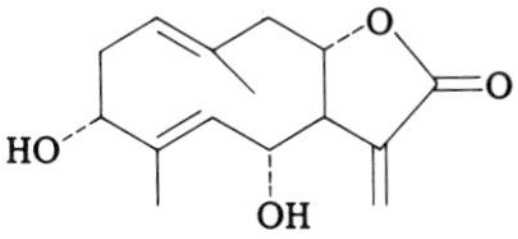

Chamissonin
C-6-α; C-8-α
24

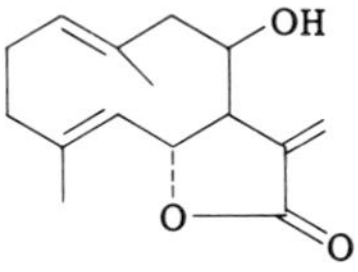

Eupatolide
C-6-α; C-8-β
25

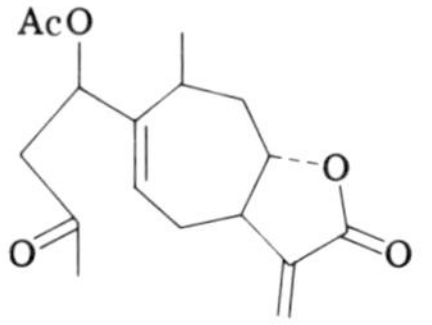

Xanthinin
C-8-α

4

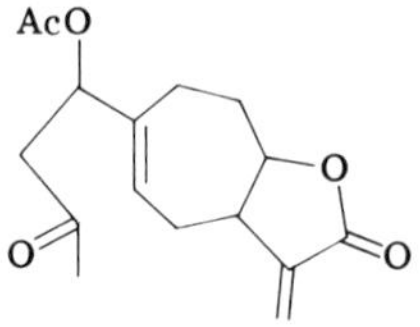

Xanthumin
C-8-β

26

It is not possible to reach a satisfactory decision as to the point in the sequence of events leading from **2** to **3** at which the lactonic oxygen atom is introduced. Arguments based on analogy with other compounds can be found to support the view that ring oxygenation occurs before construction of the C-12 carboxylic acid group (step A, below) or to support the alternate view that this oxygenation is the later process (step B, below):

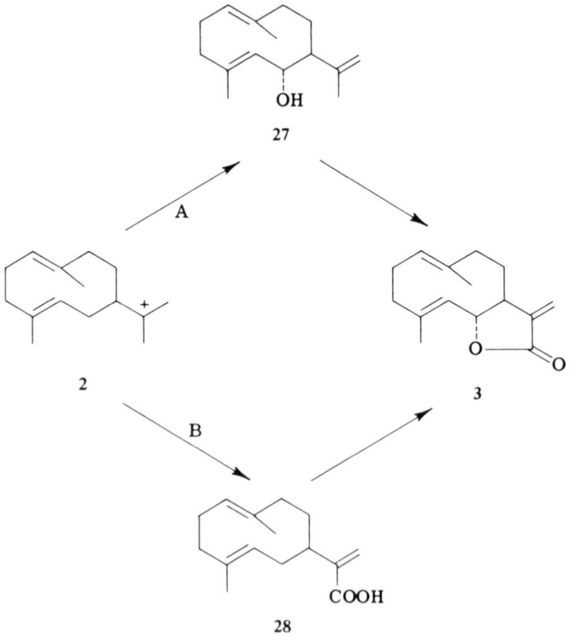

27

A

2

3

B

28

That course **2** → **27** → **3** (or its counterpart in a bicyclic precursor) is possible is shown by the occurrence in nature of sesquiterpenoid compounds containing C-6 or C-8 oxygen functions but with unoxidized side chains (isopropenyl groups or the equivalent). Among these are germacrone (**29**) and the recently discovered pygmol (**30**) (Irwin, 1971).

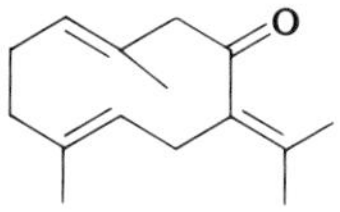

Germacrone

29

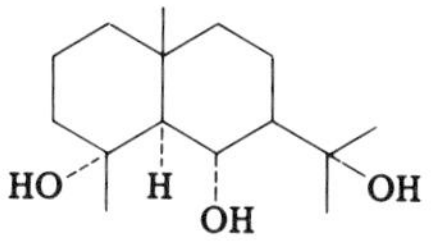

Pygmol

30

Pygmol is a constituent of an *Artemisia,* a genus rich in sesquiterpene lactones, most of them with C-6-α/C-7-β lactone fusion. It may be significant that *A. pygmea,* in which **30** occurs, does not contain lactones of this class.

On the other hand, *Ambrosia* and *Hymenoclea* species, as well as *Artemisia vachanica,* contain ilicic acid (**31**), suggesting that a course **2** → **28** → **3** is possible. Occurring along with ilicic acid (cf. costic acid) are a host of sesquiterpene lactones (with both C-6/C-7 and C-7/C-8 lactone closures).

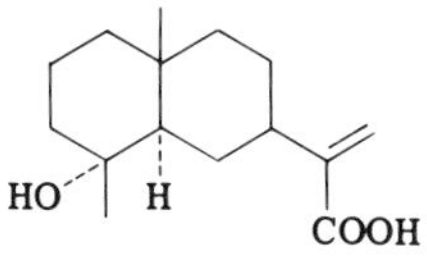

Ilicic acid

31

A suggestion of the relationship between lactone formation and carbon-hydroxylation is found in an examination of the lactones that occur in various genera of the tribe Ambrosiae, and in particular in species of *Ambrosia.* The typical compounds of this group are C-6/C-7 cis-fused lactones, as in ambrosin (**6**); but C-7/C-8 lactones (both cis and trans) also occur in the class. An illuminating example is found in the species *Ambrosia chamissonis,* two morphologically distinct (extreme) forms of which are known. One of these is characterized by the presence of chamissonin (**24**) as the principal constituent; the other, by costunolide (**3**). Although the possibility cannot be excluded, it appears unlikely that costunolide is the immediate precursor of chamissonin, for this would require not only hydroxylation of costunolide at C-3 and C-8, but also the shifting of the lactone ring from C-6/C-7 to C-7/C-8. It seems much more likely that hydroxylation occurs as in the following scheme:

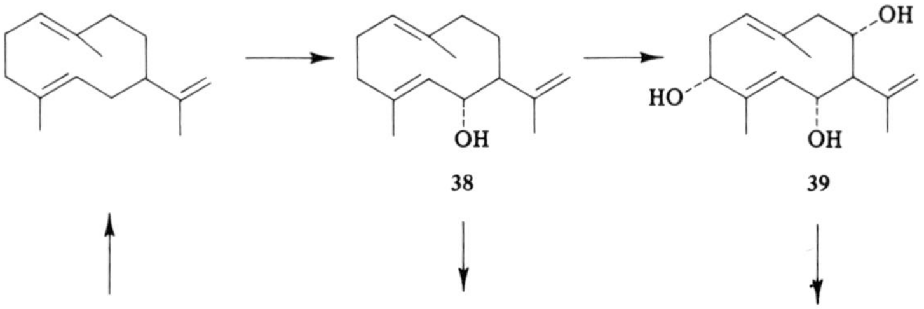

38 39

2 3 24

Since ilicic acid (**31**) is a component of *Hymenoclea*, a genus closely allied to *Ambrosia*, it is possible to modify the foregoing scheme to the following:

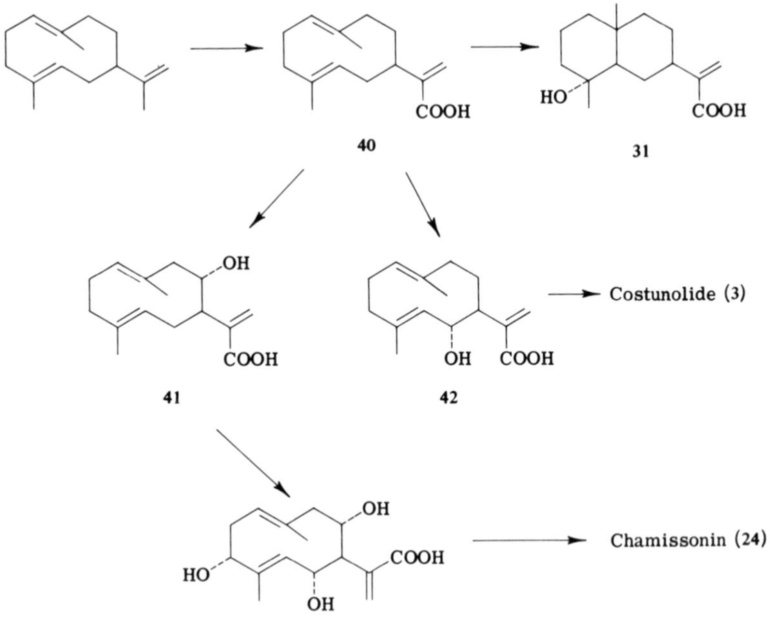

40 31 41 42 43

There is, however, no persuasive basis for excluding the possibility that C-6 or C-8 hydroxylation can occur both before and after oxidation of the side chain to the carboxylic acid.

The manner in which the C-6 or C-8 oxygen atoms are introduced is not known, but ample evidence exists that direct oxidation of C—H to C–OH is a common biological process. Experiments with the sesquiterpenoid compound guaioxide (**32**) have shown that this compound can be oxidized upon incubation with the microorganism *Mucor parasiticus* (Takeda, 1970); the products **33–37** have been found:

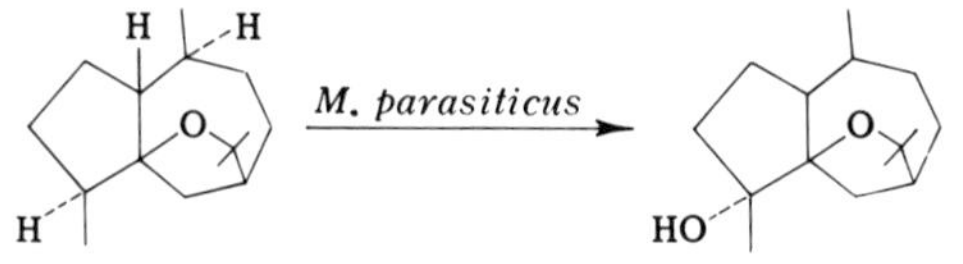

M. parasiticus

Guaioxide **32**   **33**   **34**

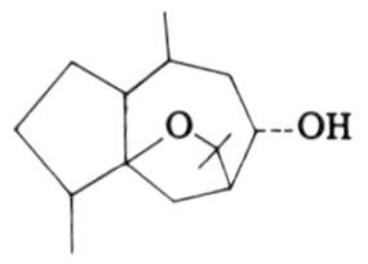

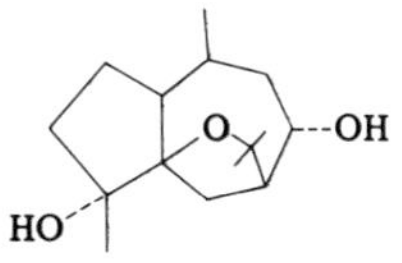

35 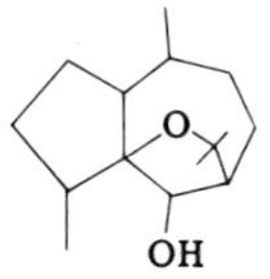   **36** 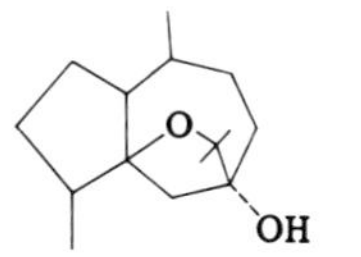   **37**

Eleven other microorganisms were shown to be able to produce two or more of the oxidation products **33–37** when incubated with **32**.

None of the above considerations provides for persuasive conjecture. The discussion to follow will embrace further alteration of compounds in which the lactone ring is present.

Oxidative Formation of a Primary Lactonic Precursor

The precursor **2** is related by a straightforward acid-catalyzed cyclization to the naturally occurring cryptomeridiol (**44**):

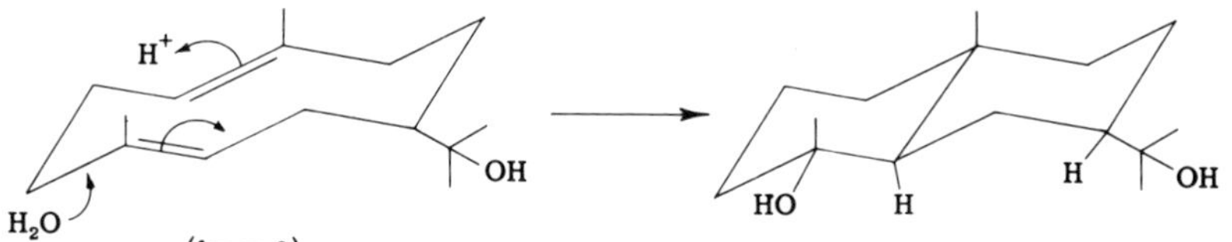

(from 2)   Cryptomeridiol **44**

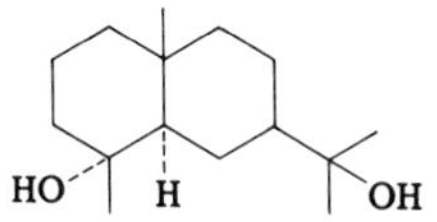

44

Oxidation of **44** at C-6 yields pygmol (**30**).

The counterpart of this reaction in the lactone series can be written starting with costunolide (**3**). The product of this cyclization is the lactone arbusculin A (**45**), a constituent of *Artemisia arbuscula* Nutt (Irwin, 1971).

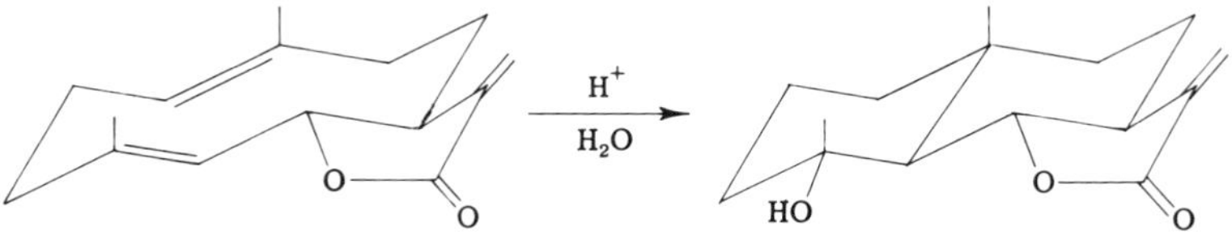

H^+ / H_2O

Costunolide

3

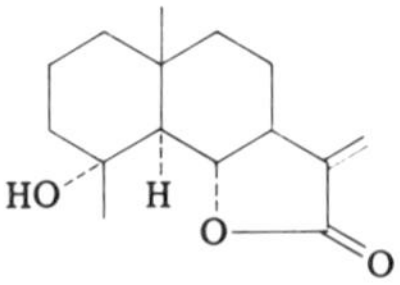

Arbusculin A

45

Arbusculin B (**46**) ,also found in *A. arbuscula*, possesses the $\Delta^{4,5}$ double bond. It is formed as one of the three possible products by dehydration (*in vitro*) of arbusculin A*:

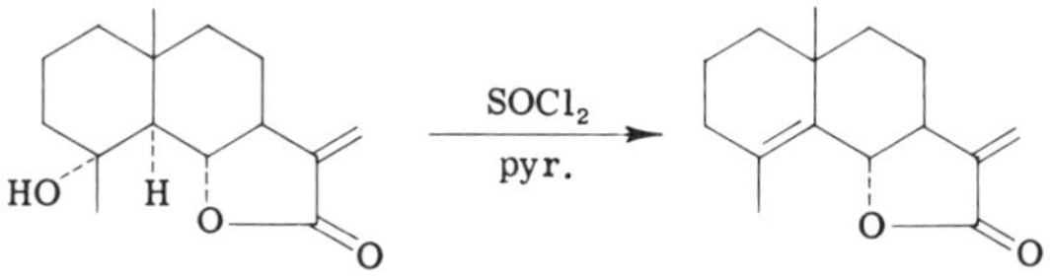

$SOCl_2$ / pyr.

Arbusculin A

45

Arbusculin B

46

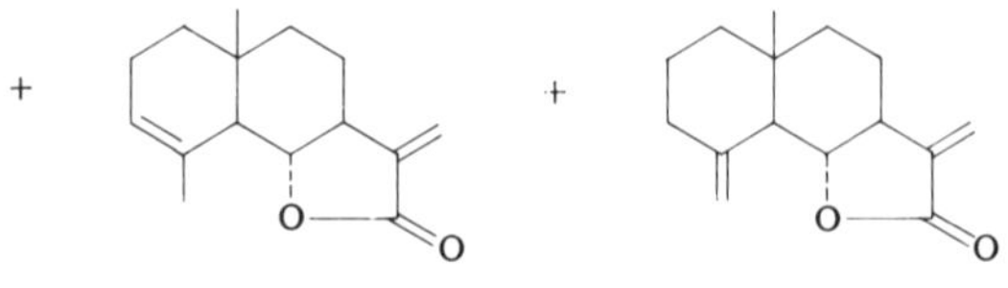

\+ **47** + **48**

* Although **47** and **48** have not been found in the plant along with **45** and **46**, the caution must be noted that failure to isolate a compound is not necessarily proof of its absence in the plant.

It is known from experiments of this kind with other 4-β-CH_3/4-α-OH/5-α-H lactones that dehydration leads predominantly to the exocyclic methylene compound (such as **48**). That arbusculin B (**46**) is produced in the laboratory in only minor amount, and yet is the only one of **46**–**48** detected in the plant, suggests strongly that it is not formed from **45**, but is a direct product of the original cyclization process. It is equally improbable that **46** is formed from **3** in a concerted protonation (at C-1)–deprotonation (at C-5) process, for this kind of reaction would be expected to lead to the $\Delta^{3,4}$ or $\Delta^{4,14}$ product, both of which types are of wide occurrence among the sesquiterpene lactones of the santanolide class. The intermediacy of an (enzymatically stabilized) carbonium ion seems probable:

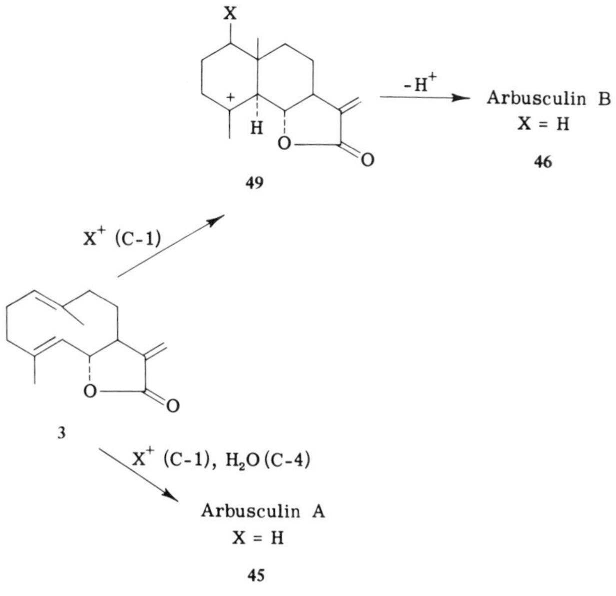

X
H
O
O
49
$-H^+$
Arbusculin B
X = H
46
X^+ (C-1)
O
O
3
X^+ (C-1), H_2O (C-4)
Arbusculin A
X = H
45

11,13-Dihydroarbusculin A (**50**) is not found in *A. arbuscula*, but occurs naturally as a constituent of the closely allied species *A. tripartita* ssp. *rupicola* (Irwin, 1971). Whether it is formed from a dihydrocostunolide (or other germacranolide precursor), or by recuction of its 11,13-unsaturated analog **45** is now known. It does not appear that surmise on this point would be fruitful in the absence of additional information. It is to be noted that numerous examples are known of the occurrence in a single plant of 11-methylene and 11-methyl pairs of lactones.

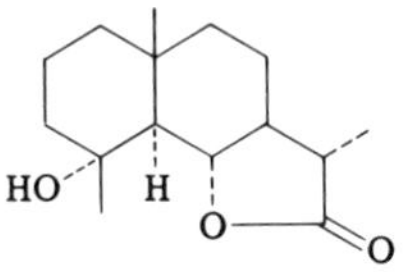

11,13-Dihydroarbusculin A

50

Elaboration to Higher Oxidation States

The discussion in the following pages will be seen to be developed from a central hypothesis—namely, that oxidative elaboration of the sesquiterpene lactones, both in ring closure of cyclodecadienolides to bicyclic compounds, and in later introduction of oxygen, is often the consequence of an initial epoxidation of a carbon-carbon double bond followed by acid-catalyzed transformations.

A direct counterpart of the electrophilic (by proton attack) ring closure **3** → **45** is found in reactions leading to compounds possessing a hydroxyl group at C-1. The process can be described in empirical terms by substituting the hypothetical OH^+ for H^+ in the ring-closure step. It is probable that the biological process is the acid-catalyzed attack upon an epoxide:

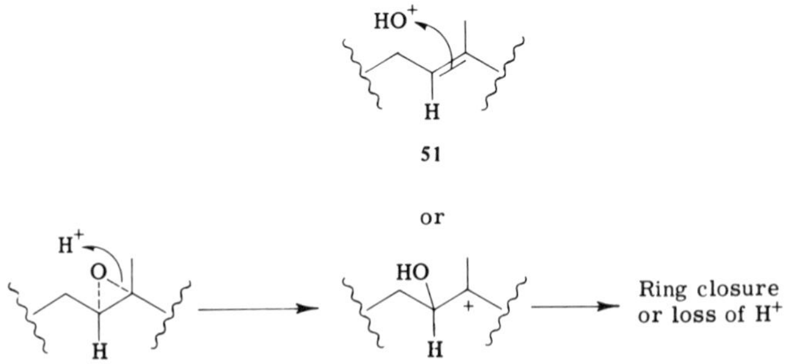

51

or

Ring closure or loss of H^+

52 53

Two compounds, douglanine (**54**) (Matsueda and Geissman, 1967a) and santamarine (**55**) (Romo de Vivar and Jimenez, 1965), both found in *Artemisia* species (tribe Anthemideae), are known, differing in that one is α-C-1-OH, the other β-C-1-OH:

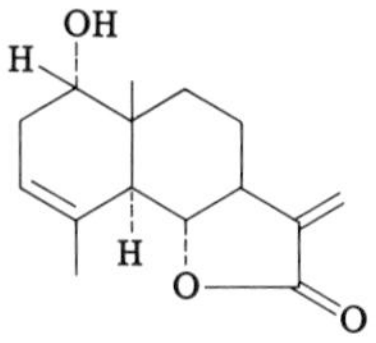

Douglanine

54

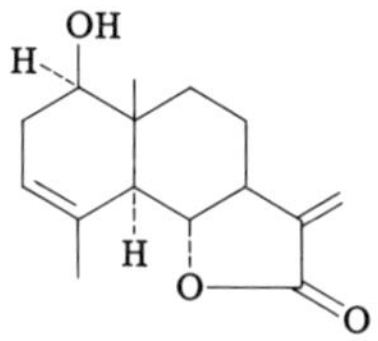

Santamarine

55

The formation of santamarine (**55**) can be formulated as proceeding by way of the series **52** → **53**:

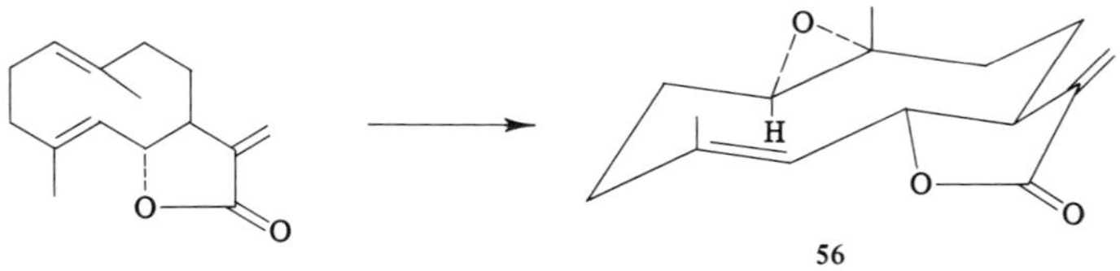

3

56

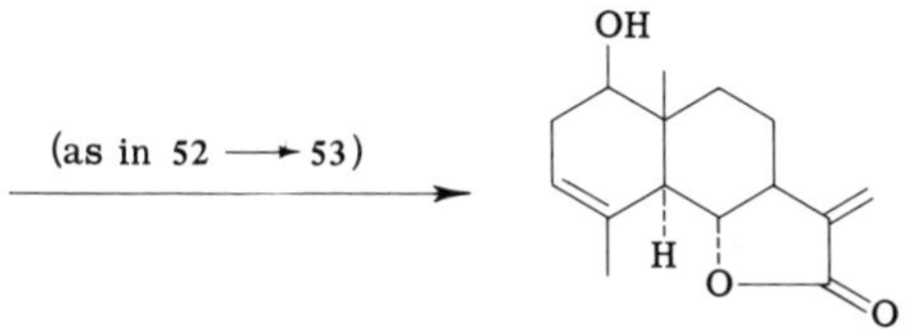

(as in 52 → 53)

Santamarine

55

Douglanine (**54**), however, cannot be formed by the same concerted ring closure of the oxide **56** (or by the equivalent process **51** → **53**), and thus must be the result of an alternate route. A clue is found in the following observations.

Artemisia douglasiana, from which douglanine is isolated, is a member of a group of species which are closely allied by morphology, most of which possess synonymous names as variations or subspecies of the species *vulgaris*. The study of several species of the *Vulgaris* complex has provided information which constitutes persuasive evidence for biosynthetic pathways. *Artemisia verlotorum* Lamotte* contains the compounds artemorin (**57**), verlotorin (**58**) and dehydroartemorin (**59**) (Geissman, 1970).

* This has been found to comprise (at least) two distinct chemovariant forms, one of which contains vulgarin (**68**), the other artemorin (**57**) and its allies.

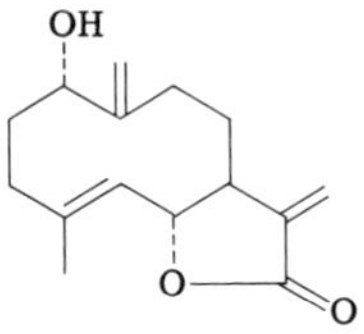

Artemorin

57

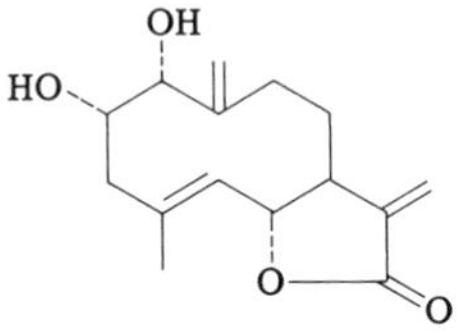

Verlotorin

58

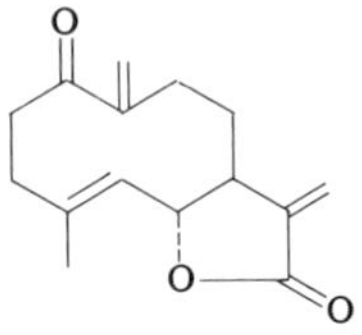

Dehydroartemorin

59

It appears to be significant that artemorin is the C-1 α-hydroxy compound. Ring closure of artemorin by the following acid-catalyzed process can lead directly to douglanine (**54**), the C-1 epimer of **55**:

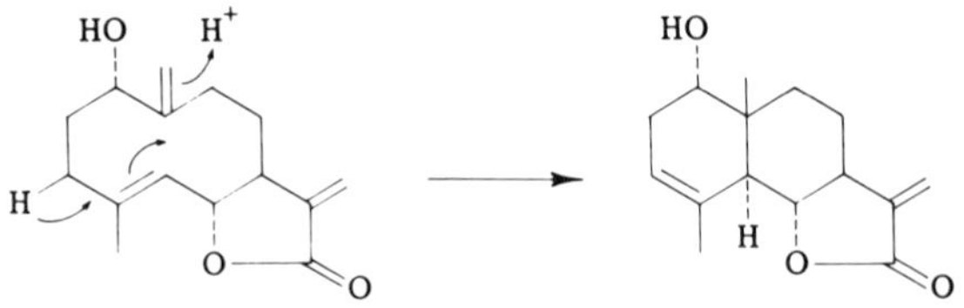

57

54

The biosynthetic difference in the formation of douglanine and santamarine thus appears to lie in a different mode of initial epoxidation. The epoxide leading to **57**, which is not disposed to cyclize by the route **56** → **55**, undergoes the alternate acid-catalyzed course in which epoxide (**60**) opening leads directly to **57**:

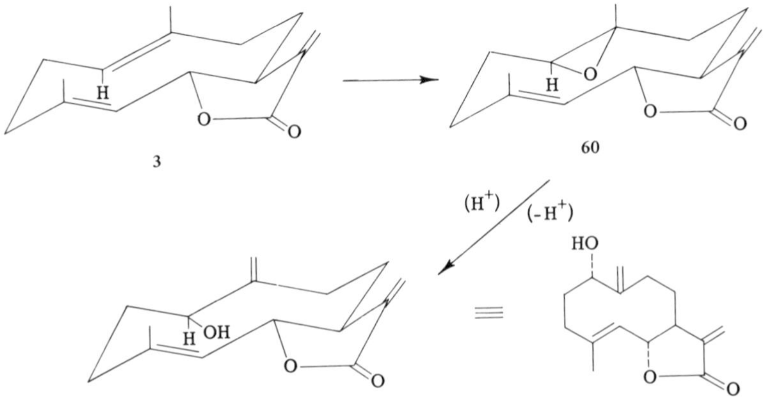

3

60

(H^+) $(-H^+)$

57

57

A related case has been observed in a pair of compounds isolated from

Artemisia tridentata ssp. *tridentata* (Irwin, 1971). These are dentatin A (**61**) and B (**62**):

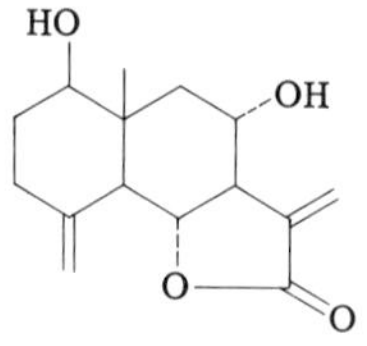

Dentatin A

61

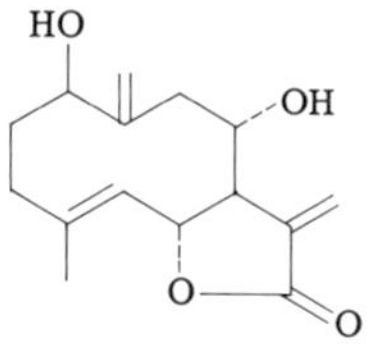

Dentatin B

62

It would appear that dentatin B is formed by an alternate mode of opening of an 8-hydroxy epoxide corresponding to **56** (which in that case leads to santamarine), leading here to dentatin B. From these observations the suggestion may be made that *Chrysanthemum* is more closely related to the Seriphidium members of *Artemisia* than to those classed as the section Abrotanum (e.g., *A. douglasiana*). This view is supported by the occurrence in *Achillea* and *Matricaria*, close allies of *Chrysanthemum*, of compounds of the matricarin (**63**) group, which are absent from species of the section to which the *Vulgaris* complex belongs but are typical of a number of species of the Serephidium section.

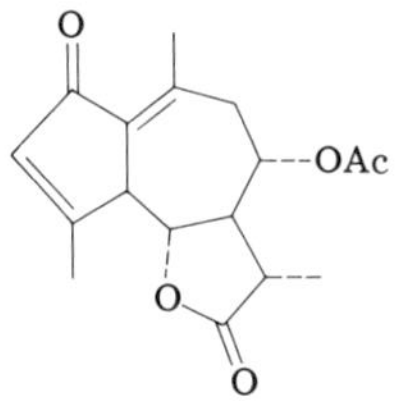

Matricarin

63

The importance of epoxidation as a process by which oxidative elaboration of sesquiterpenes occurs can be further illustrated by other constituents of members of the *Vulgaris* complex. *A. ludoviciana* contains the compounds, **54**, **64**, **65**, **66**, and **67**, shown in Scheme 1 (Lee and Geissman, 1970). Compound **54** is found also in *A. douglasiana*, and **57** in *A. verlotorum*. It will be apparent that these lactones can be put into a progressive sequence of structural elaboration; the starting cyclodecadienolide, **3**, has not, however, been found in these plants. A constituent of *A. verlotorum** that appears to be the counterpart of **65** (Matsueda and Geissman, 1967b) in the 11,13-dihydro series is vulgarin (**68**) (Geissman and Ellestad, 1962). No 11,13-dihydro derivatives of **54**, **64**, **66**, or **67** have, however, been found.

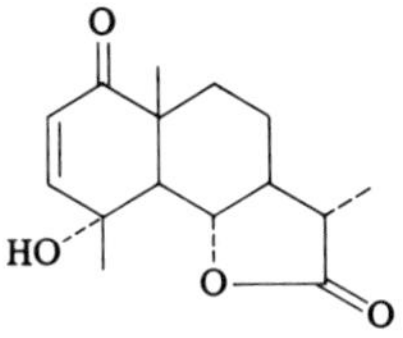

Vulgarin

68

The question of whether vulgarin (**68**) arises by reduction of arglanine (**65**) (at 11,13), or by way of a 11,13-dihydro series corresponding to Scheme 1 cannot be answered at this time.

Certain variants of the above schemes can be discerned in other compounds of *Artemisia* species. A group of lactones from *A. arbuscula* include the compounds **45**, **46**, **69**, **70**, and **71** (Irwin, 1971).

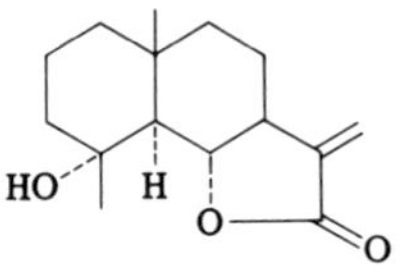

Arbusculin A

45

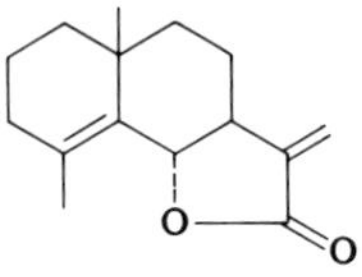

Arbusculin B

46

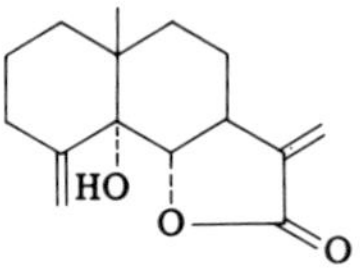

Arbusculin C

47

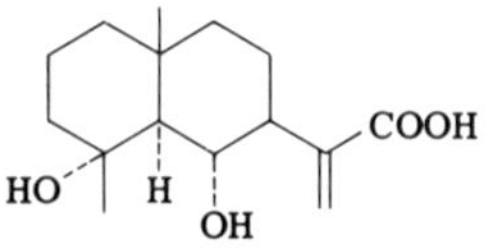

Arbusculin E

70

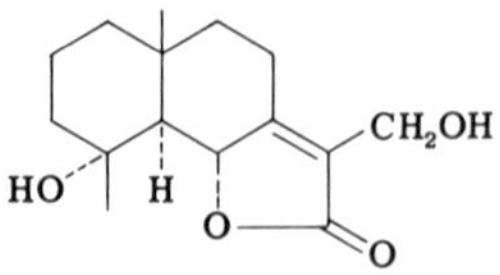

Arbusculin D

71

* Two distinct chemovars of *A. verlotorum* have been found. One contains artemorin (**57**) and the related verlotorin (**58**) and anhydroverlotorin (**59**). The other (morphologically identical with the first) contains none of these, but contains dihydro-**65** [vulgarin (**68**) = 11,13-dihydroarglanine].

Scheme 1

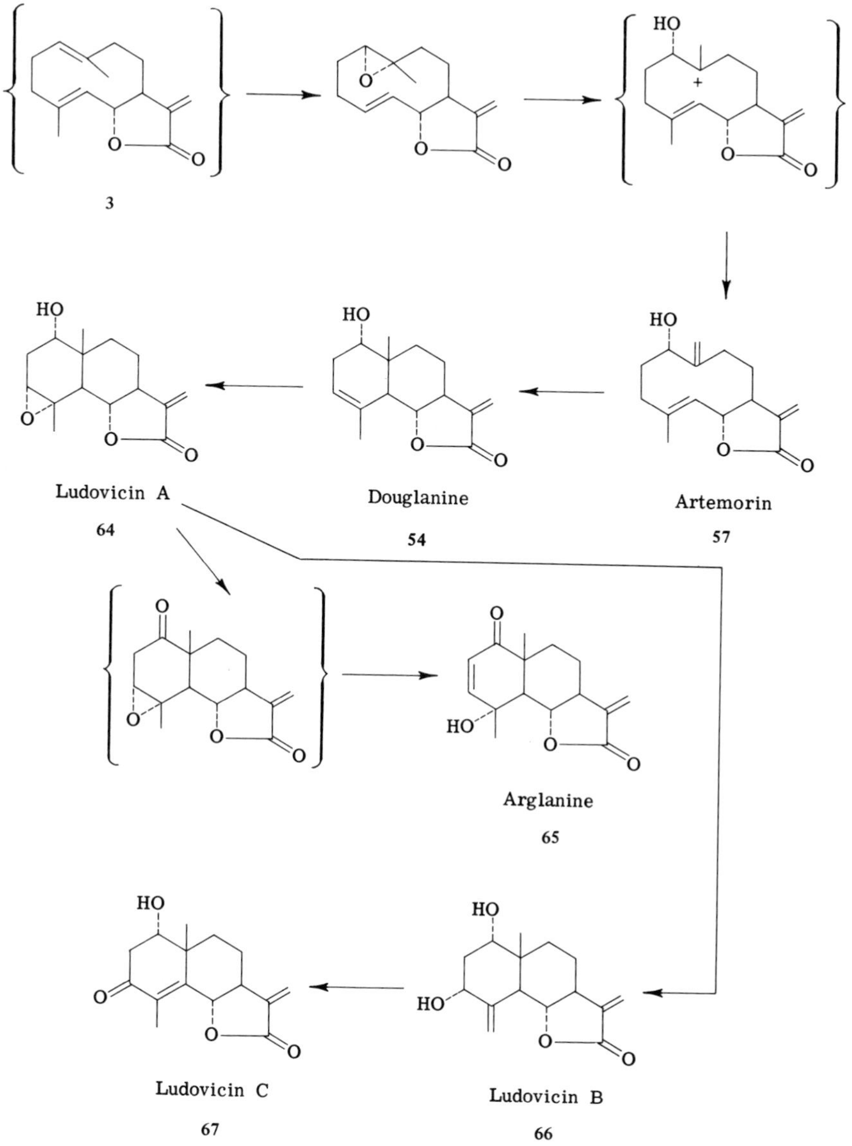

3

Ludovicin A

64

Douglanine

54

Artemorin

57

Arglanine

65

Ludovicin C

67

Ludovicin B

66

The derivation of arbusculin A and B from the precursor **3** by non-oxidative acid-catalyzed ring closures has been described above. The formation of arbusculin C (which has been accomplished in the laboratory) by epoxidation of arbusculin B and subsequent ring opening can be represented in the following way:

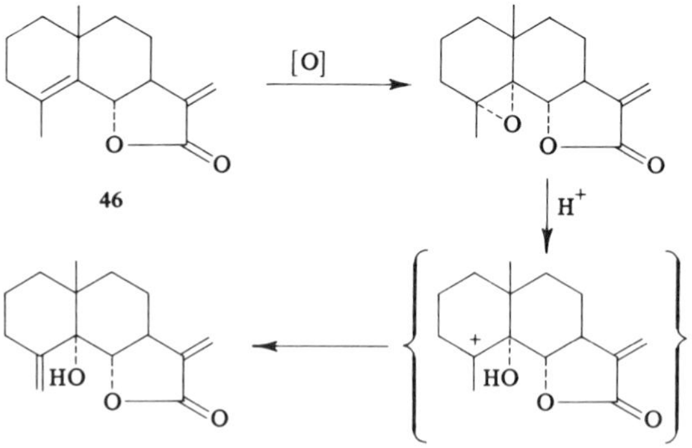

[O]

46

H^+

Arbusculin C

69

A comparable pair of compounds in *A. rothrockii* bear the same relationship; these are rothin A (**72**) and B (**73**) (Irwin, 1971).

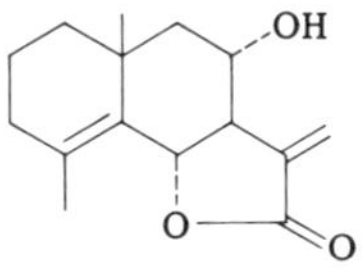

Rothin A

72

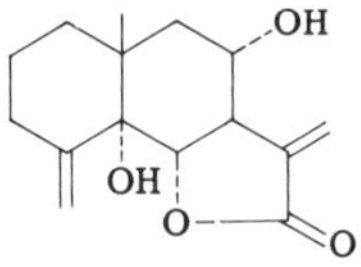

Rothin B

73

The change, artemorin (**59**) → douglanine (**54**), postulated as a step in the biosynthesis of the latter, finds a direct counterpart in ridentin (**74**) and ridentin B (**75**), constituents of *A. tridentata* ssp. *tridentata* (Irwin, 1971):

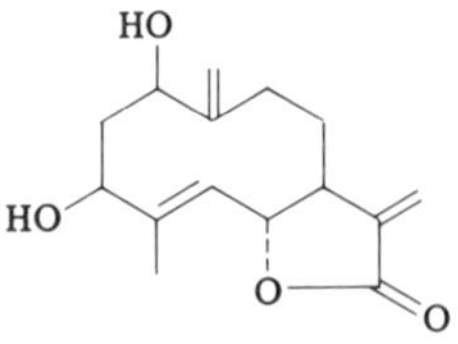

Ridentin

74

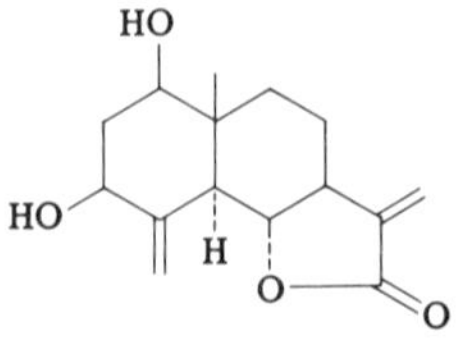

Ridentin B

75

It is further to be noted that the ketolactone artecalin (**76**), found in *Artemisia californica*, is related by an oxidation-reduction (or by prototropy) to ridentin B (**75**). The naturally occurring tuberiferine (**77**) is identical with 1,2-anhydroartecalin.

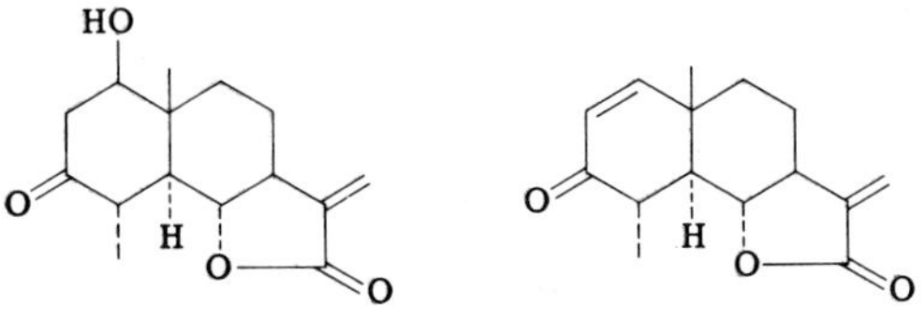

Artecalin

76

Tuberiferine

77

It will be seen that the introduction of a 3-hydroxy group, as in the ridentins or the 3-keto group of artecalin, can be accomplished by ring opening of a 3,4-epoxide:

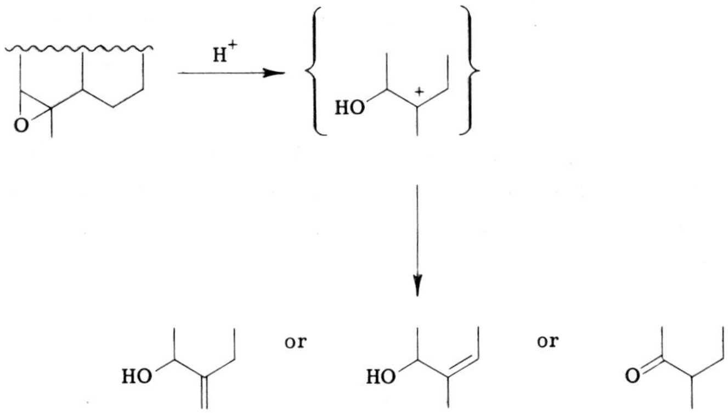

However, 3-hydroxygermacranolides (as **74**) would then be regarded as arising from a $\Delta^{3,4}$-germacranolide, but this structure is not observed to co-occur with the 3-hydroxy compounds.

Further indications of the intermediacy of an epoxide in oxidative transformations can be recognized in a group of guaianolides found in several species of *Artemisia* of the Tridentatae section. Typical of these is cumambrin B (**78**) (which is accompanied in the plants by the 8-acetate and the 8-deoxy compound). The lactones rupicolin A (**79**) and B (**80**) can be formed from **78** by C-1/C-10 dehydration, C-1/C-10 epoxidation, and opening of the epoxide ring:

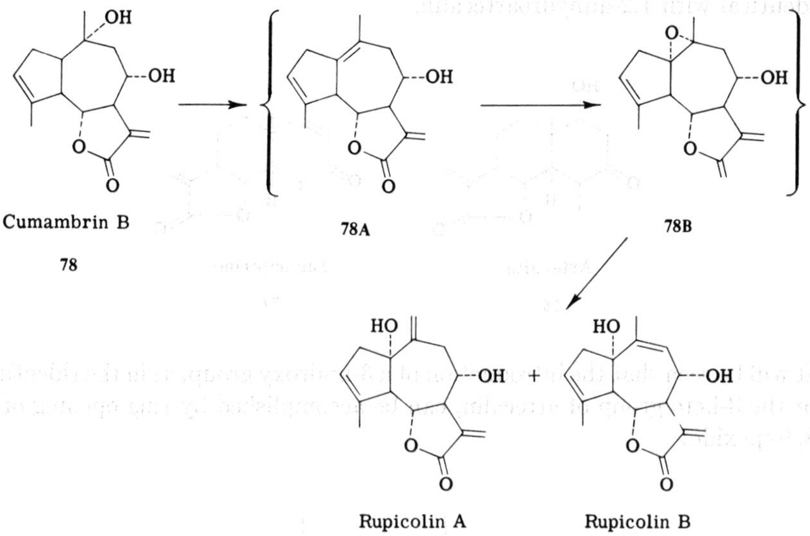

Cumambrin B

78

78A

78B

Rupicolin A

79

Rupicolin B

80

The diepoxides rupin A (**81**), rupin B (**82**) and canin (**83**), which occur in closely related species of the Tridentatae, can be seen to be derivable from **79** or **80** by mechanistically unexceptional steps, of which epoxidation is the last:

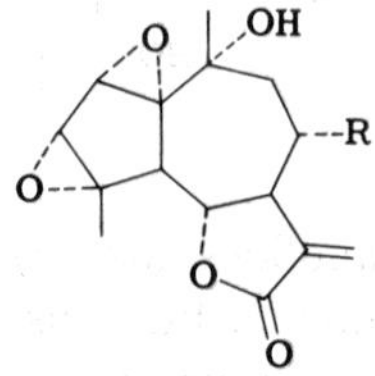

81. R = OH
82. R = OAc
83. R = H

Finally, the ubiquitous (in several species of the Tridentatae) matricarin

(63) and its congeners (**84**, **85**)

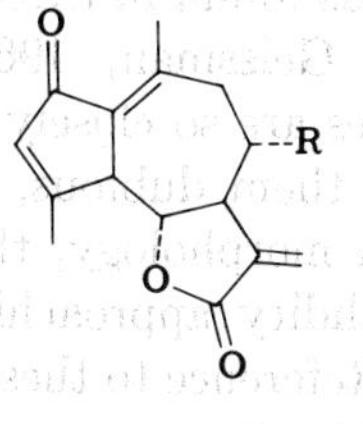

63. Matricarin; R = OAc
84. Deacetylmatricarin; R = OH
85. Deacetoxymatricarin; R = H

can be derived readily by a simple acid-catalyzed rearrangement of the oxide (**78B**) derived from the anhydrocumambrin (**78A**). Although there is no direct evidence that epoxidation is involved in the processes whereby **63**, **84**, and **85** are formed from a less highly oxidized precursor (such as **78** or 8-deoxy-**78**), their occurrence along with **78–83** in plants distinguished only by trivial morphological differences strongly suggests the intervention of epoxidation in transformations that lead to highly oxidized (oxygenated or unsaturated) end products.

Rearrangements to Pseudoguaianolide Structures

The pseudoguaianolides (ambrosanolides) are characterized by a non-regular isoprenoid skeleton. Since they carry an oxygen function at C-4, it is probable that they arise by a rearrangement of the following kind:

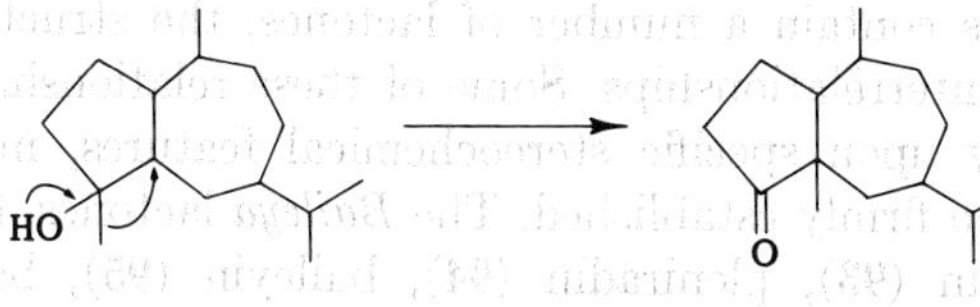

86 87

The cationic center that initiates this rearrangement could in principle be located at C-5, C-1, or C-10:

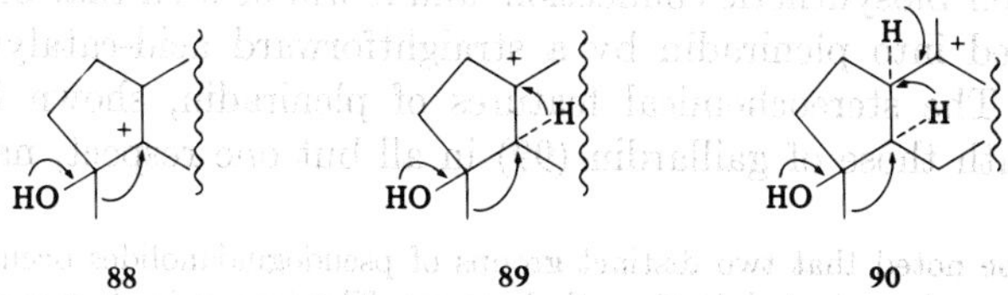

88 89 90

Indications of one possible course for this kind of rearrangement* are found in a group of compounds found in three species (of the Heliantheae) of *Baileya* (Yoshitake and Geissman, 1969; Waddell and Geissman, 1969a). Since the three species are so closely related as to make the taxonomic separations of two of them dubious, and to distinguish the third only by trivial differences in morphology, the comparison of compounds found in all three has a validity approaching that associated with co-occurrence in a single plant. Reference to these will accordingly be made by the generic name only.

Baileya contains the first discovered sesquiterpene glycoside, paucin (**91**) (Waddell and Geissman, 1969b). It is apparent that paucin can represent the prototype of the precursor of the large number of compounds containing the 2-ene-4-one structure in the 5-membered ring, for simple elimination of the element of glucose (or of water from the aglucon) gives the unsaturated ketone; in the case of paucin itself the product is aromatin (**92**), which is found in other plants of the same tribe (Helenieae) (Romo and Joseph-Nathan, 1964):

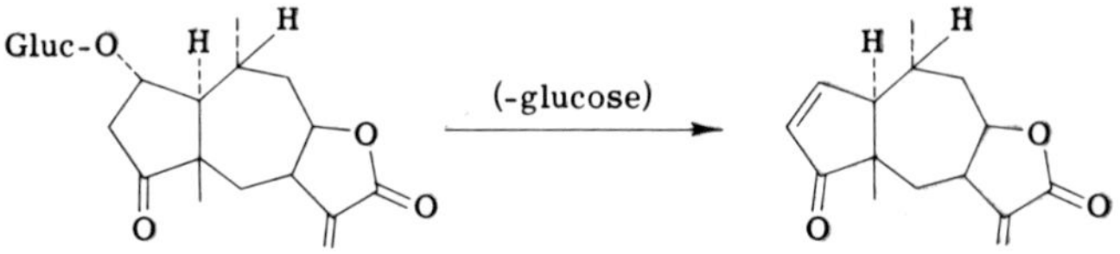

(-glucose)

Paucin
91

Aromatin
92

Baileya species contain a number of lactones, the structures of which suggest certain interrelationships. Some of these relationships depend for their plausibility upon specific stereochemical features, many of which have not yet been firmly established. The *Baileya* lactones, besides paucin (**91**), are radiatin (**93**), pleniradin (**94**), baileyin (**95**), baileyolin (**96**), plenolin (**97**), and fastigilin C (**98**) (Herz *et al.*, 1966).

Baileyin, pleniradin, and radiatin are, respectively, a germacranolide, a guanianolide, and a pseudoguaianolide. Their co-occurrence strongly suggests their biosynthetic connection, and it will be seen that baileyin can be converted into pleniradin by a straightforward acid-catalyzed transformation. The stereochemical features of pleniradin, shown in **94**, are identical with those of gaillardin (**99**) in all but one respect, namely, the

* It is to be noted that two distinct groups of pseudoguaianolides occur in nature: those with α-and β-oriented C-10 methyl groups. The former is characteristic of the tribe Helenieae; the latter, of Ambrosiae.

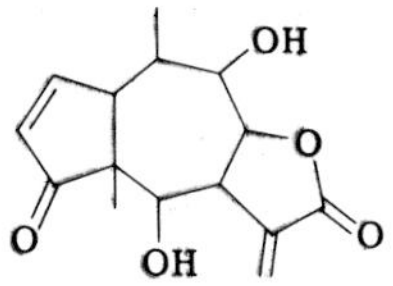

Radiatin

93

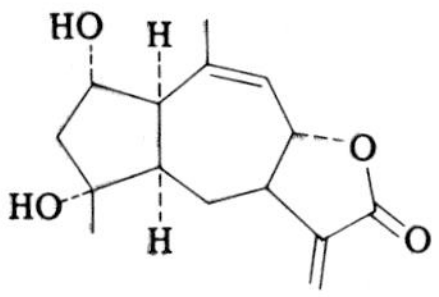

Pleniradin

94

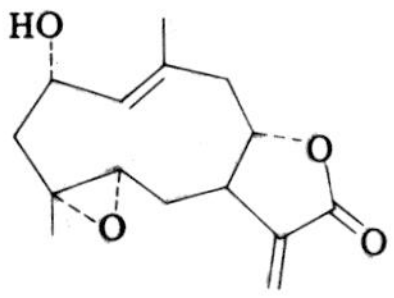

Baileyin

95

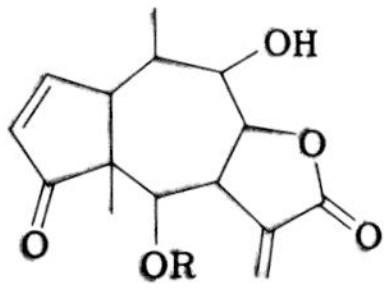

Baileyolin
OR = angeloyl

96

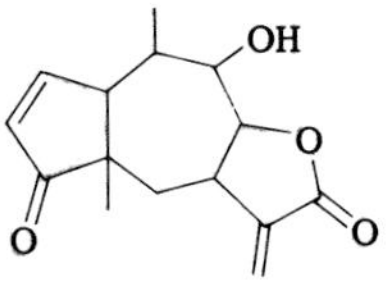

Plenolin

97

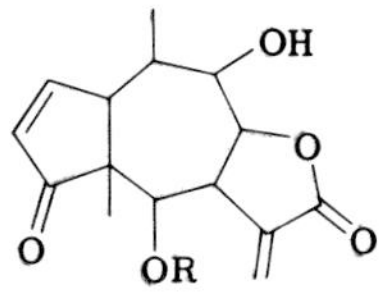

Fastigilin-C
R = senecioyl

98

configuration at C-1. Gaillardin has been subjected to X-ray analysis and has been formulated as having β-C-1-H (Dullforce *et al.*, 1969).

The conversion of pleniradin into (a) a pseudoguaianolide of the aromatin or paucin type, or (b) a 9-hydroxy pseudoguaianolide of the plenolin or fastigilin type can be formulated as follows*:

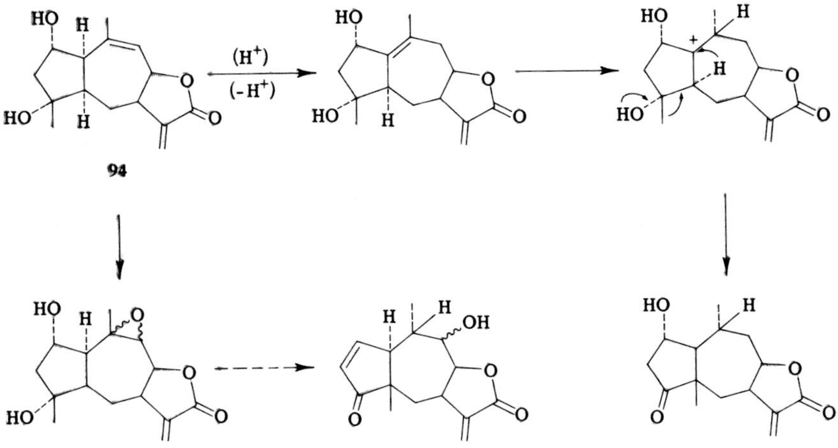

(H^+)
$(-H^+)$

94

e.g., 97

Aglucone of 91

* The configuration of the lactone is not specified, for both cis- and trans-fused C-7/C-8 lactones may be found in the same plant; e.g., aromatin and aromaticin.

Certain configurational features of these compounds are not yet established; e.g., the C-10 methyl group, usually α-disposed in the pseudoguaianolides of the Helenieae, is not known to be so oriented in **93**, **96**, and **97** (although it is α- in paucin); and the 9-hydroxyl group in **93**, **96**, and **97** is of still unknown configuration.

Gaillardin (**99**) ,which is found in a genus (*Gaillardia*) that also contains the fastigilins (e.g., **98**), has the following configuration:

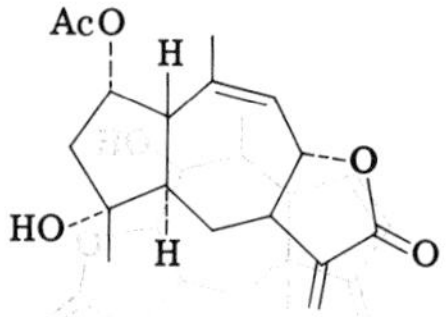

Gaillardin

99

It is apparent that its conversion into a 10-α-methyl-9-hydroxypseudoguaianolide (**100**) could proceed by a completely concerted process if it be assumed that an α-oxide at C-9/C-10 is first formed:

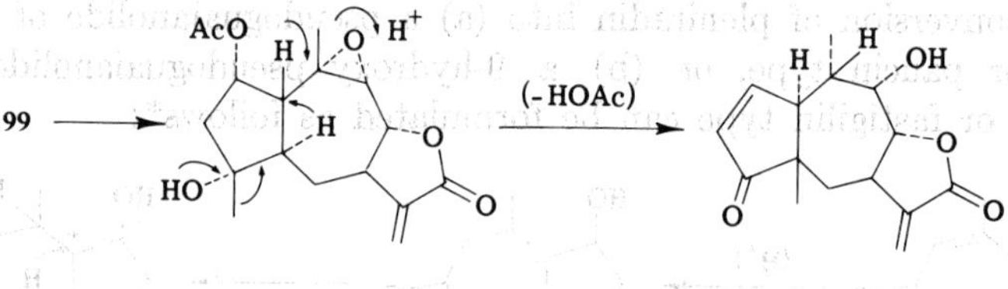

100

The prediction from this conjecture would be that the hydroxyl group at C-9 in **93**, etc., is α-disposed. That this is so is not yet known with assurance. It is of interest to note that when gaillardin is brominated, the resulting 9-bromo isogaillardin (with a C-10/C-1 double bond) is indeed the α-bromo compound **101**:

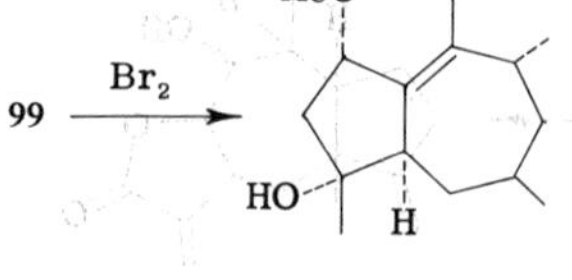

101

Pleniradin has been assigned the stereochemistry shown in **94** with the C-7/C-8 lactone trans-fused. Gaillardin (**99**) and pleniradin acetate are not the same; thus, pleniradin is assigned the C-1 α-H configuration since gaillardin has been shown by X-ray crystallographic evidence to have C-1 β-H.

There appears to be no absolute necessity for a single stereochemical course for germacranolide → guaianolide ring closure nor for guaianolide → pseudoguaianolide rearrangement, for the pseudoguaianolides of the Ambrosieae have C-10 β-methyl groups, while those of the Helenieae have C-10 α-methyl groups (so far as is known).

The formation of xanthanolides, e.g., xanthinin (**4**), and xanthumin, its 8-epimer, can be rationalized by postulating another course for the rearrangement of the intermediate cation (**102**) derived from **94***:

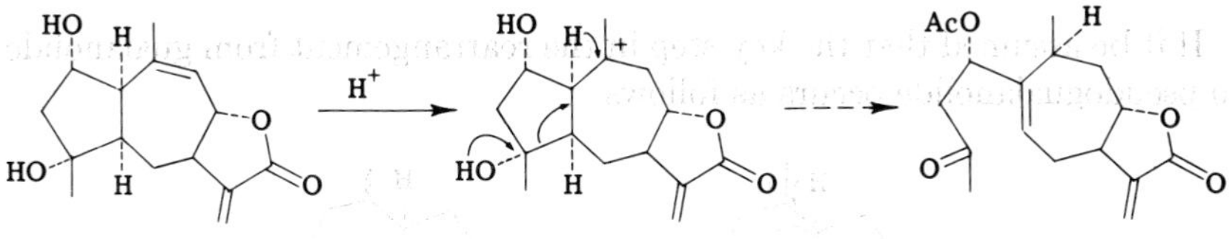

94 102 4

Stereochemical Conditions for C-4 Methyl→C-5 Methyl Rearrangement

A consideration of the structures of the known pseudoguaianolides reveals that there are certain restrictions upon the number of possible stereoisomers with respect to (a) configuration at C-10 and (b) the position (C-6/C-7 or C-7/C-8) and mode (cis or trans) of lactone closure. The following represents the structural types presently known (only the configurations at C-10, C-6, and C-8 are relevant to this discussion). Also shown are the (still?) unknown configurational types (**106–108**).

Known

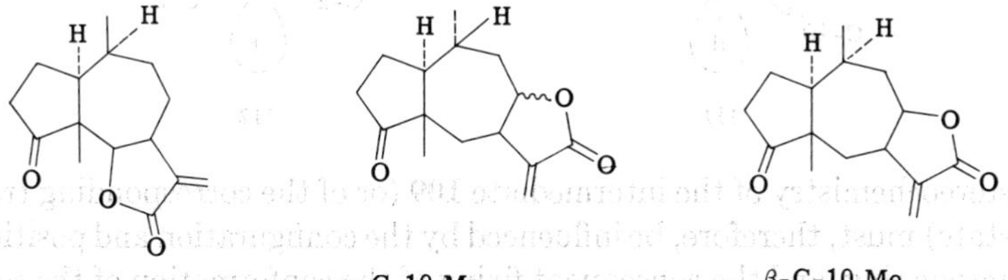

β-C-10 Me
C-6/C-7 *cis*

103

Known

α-C-10 Me
C-7/C-8 *cis* and *trans*

104

Known

β-C-10 Me
C-8/C-8 *cis*

105

* The configuration of the —OAc group at C-2 of **4** is not known with certainty. The course **94** → **102** → **4** would require that it have the configuration shown here, but the true precursor could be other than **94** itself.

Unknown

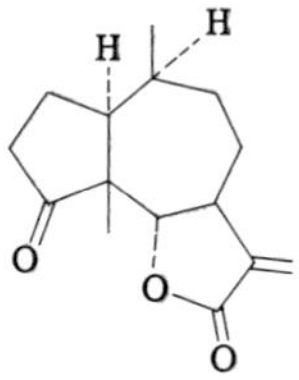

β-C-10 Me
C-6/C-7 *trans*

106

Unknown

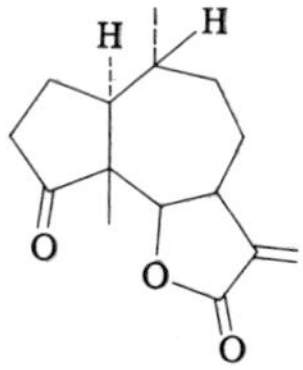

α-C-10 Me
C-6/C-1 *cis*

107

Unknown

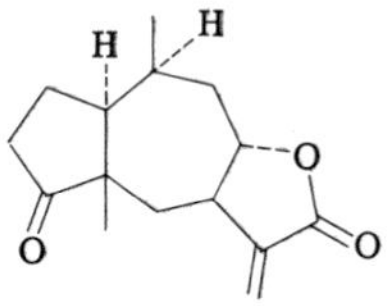

β-C-10 Me
C-7/C-8 *trans*

108

If it be assumed that the key step in the rearrangement from guaianolide to pseudoguaianolide occurs as follows

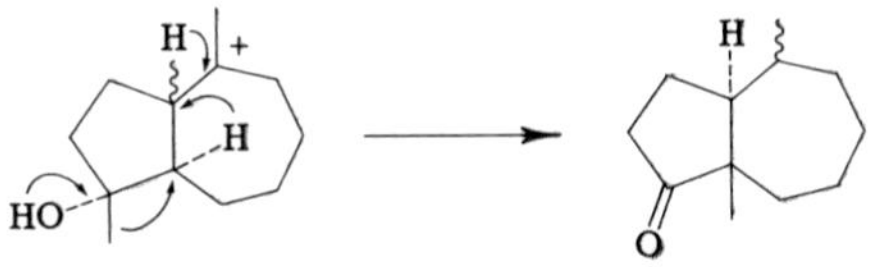

109 → **110**

the occurrence of this step, to lead to **103–105**, and its nonoccurrence to lead to **106–108** can be attributed to conformational influences. In short, if the conformation is **111** or **112**, rearrangement will occur; if **113** or **114**, it will not occur:

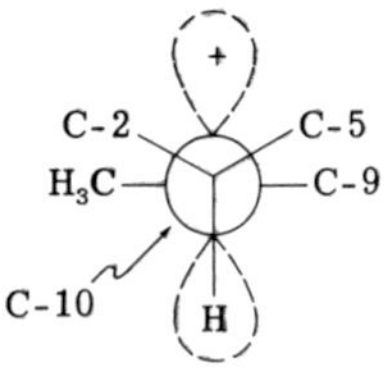

111 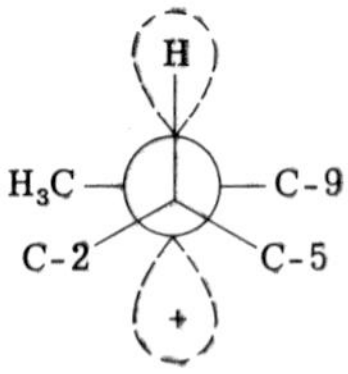 **112**

The stereochemistry of the intermediate **109** (or of the corresponding transition state) must, therefore, be influenced by the configuration and position of the lactone ring, and the consequent fixing of the conformation of the seven-membered ring (and the C-1/C-5 ring junction) in a favorable (**111, 112**) or unfavorable (**113, 114**) disposition. Although the influences cannot be persuasively shown by abstract considerations, examination of Dreiding models of **109** containing lactone ring fusions of the various types does indeed show that the expected required conformations are favored. The

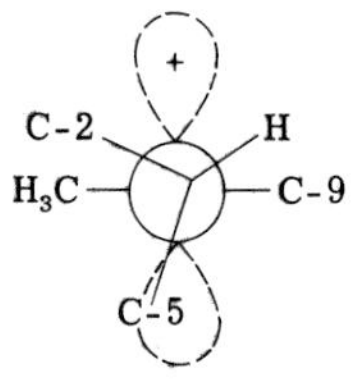

C-2 H
H_3C C-9
C-5

113

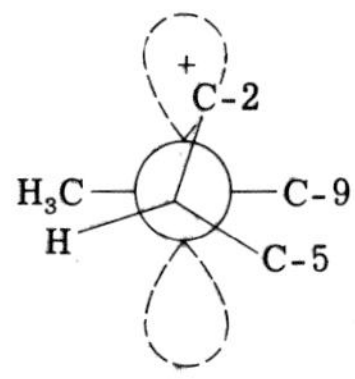

C-2
H_3C C-9
H C-5

114

quasi-chair conformations that would appear to be energetically favorable are those that meet the requirements.

The above arguments are admittedly tenuous and depend for their rationale upon the configurations of the natural lactones (**103–105**). The *caveat* must be entered that enzymatic control is undoubtedly of importance in the processes under discussion, and that the energetic considerations implied above are not necessarily controlling (Waddell, 1969).

It is significant to note that cumambrin B (**78**) is found both in species of *Ambrosia* (Ambrosieae) and *Artemisia* (Anthemideae). No ambrosanolides have been found in plants of the latter group, and no C-6/C-7 transfused pseudoguaianolides are known in the former. Cumambrin B appears to have arisen by a stereospecific oxidative ring closure of the following type:

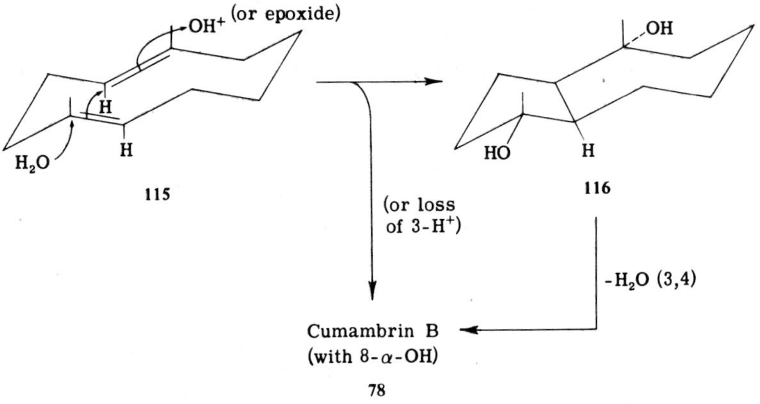

OH^+ (or epoxide)
H
H_2O H

115

OH
HO H

116

(or loss of 3-H^+)

-H_2O (3,4)

Cumambrin B (with 8-α-OH)

78

The absence of lactones of the type **106** suggests that cumambrin B (or the corresponding 3,4-hydrate, **116**) cannot undergo the transformation to the type **106**, and thus remains behind the energy barrier and is found in the plant (in *Ambrosia*) or undergoes alternate alteration (in *Artemisia*) to the more complex (more highly oxygenated) lactones characteristic of the latter genus.

Summary

A comparison of the structures of a large number of sesquiterpene lactones, restricting the comparisons to those of compounds found in the same or closely allied species, leads to rational routes of transformation of simple precursors to more complex structures. The structural hypotheses are based upon the premises that (a) an initial cyclodecadiene, formed by the cyclization of farnesol, lies at the start of the sequence of elaboration; (b) stereospecific acid-catalyzed ring closures lead to the formation of eudesmane and guaiane skeletons; (c) epoxidation of double bonds is the first step in oxidative ring closures; (d) epoxidation in later stages leads to introduction of oxygen; (e) the cyclopentenone unit of pseudoguaianolides arises from a 2-hydroxy (or glucosidoxy)-4-keto pseudoguaianolide; and (f) pseudoguaianolides arise by stereospecific rearrangements of 4-hydroxyguaianolides.

The inner consistency of the hypotheses should serve as a rational guide for the design of the experiments that will be required to place these biosynthetic processes upon a firm basis of fact.

A consideration of the available evidence shows the importance of careful examination of plants for the discovery of as many of the constituents as possible for certain aspects of the hypotheses presented suggest the occurrence of compounds that have not in fact been isolated. The accumulation of further information on the constituents of plants of closely allied botanical types can add to or refine some of the conjectures presented.

Acknowledgment

Most of the work described in this article was supported by a research grant from the U.S. Public Health Service, GM-14240. The experimental results on *Artemisia* are in large part the products of the work of Mr. M. A. Irwin. To him, and to other co-workers cited in the references, the author gives grateful acknowledgment.

References

Anchel, M., T. C., McMorris, and P. Singh. 1970. *Phytochemistry* **9**:2339.

Birch, A. J., and S. F. Hussein. 1969. *J. Chem. Soc. C* p. 1473.

Dullforce, T. A., G. A. Sim, D. N. J. White, J. E. Kelsey, and S. M. Kupchan. 1969. *Tetrahedron Lett.* p. 973.

Geissman, T. A. 1970. *Phytochemistry* **9**:2377.

Geissman, T. A., and G. A. Ellestad. 1962. *J. Org. Chem.* **27**:1855.

Hendrickson, J. B. 1959. *Tetrahedron* **7**:82.

Herz, W., S. Rajappa, S. K. Roy, J. J. Schmid, and R. N. Mirrington. 1966. *Tetrahedron* **22**:1907.

Irwin, M. A. 1971. Ph.D. Thesis, University of California, Los Angeles, California.
Jones, R. V. H., and M. D. Sutherland. 1968. *Chem. Commun.* p. 1229.
Lee, K. H., and T. A. Geissman. 1970. *Phytochemistry* **9**:403.
Matsueda, S., and T. A. Geissman. 1967a. *Tetrahedron Lett.* p. 2013.
Matsueda, S., and T. A. Geissman. 1967b. *Tetrahedron Lett.* p. 2159.
Parker, W., J. S. Roberts, and R. Ramage. 1967. *Quart. Rev.* (*Chem. Soc.*) **3**:331.
Romo, J., and P. Joseph-Nathan. 1964. *Tetrahedron* **20**:79.
Romo de Vivar, A., and H. Jimenez. 1965. *Tetrahedron* **21**:1741.
Ruzicka, L. 1963. *Pure Appl. Chem.* **6**:493.
Staunton, J. 1969. *Annu. Rep. Chem. Soc.* p. 555.
Takeda, K. 1970. *Pure Appl. Chem.* **21**:181.
Waddell, T. G. 1969. Ph.D. Thesis, University of California, Los Angeles, California.
Waddell, T. G., and T. A. Geissman. 1969a. *Phytochemistry* **8**:2371.
Waddell, T. G., and T. A. Geissman. 1969b. *Tetrahedron Lett.* p. 515.
Waldner, E. E., C. Schlatter, and H. Schmid. 1969. *Helv. Chim. Acta* **52**:15.
Yoshitake, A., and T. A. Geissman. 1969. *Phytochemistry* **8**:1753.

RECENT DEVELOPMENTS IN THE BIOSYNTHESIS OF PLANT TRITERPENES

T. W. GOODWIN

Department of Biochemistry, University of Liverpool, Liverpool, England

Introduction

Most of the work in our group over the past few years has been oriented toward elucidation of the details of triterpene and carotenoid biosynthesis with the aid of various stereospecifically labeled species of mevalonic acid. This has recently been reviewed in detail (Goodwin, 1971), so here I intend to concentrate on our more recent investigations and to discuss them in relation to the general problems of terpenoid biosynthesis.

The First Cyclic Precursor of Plant Sterols

It now appears clear that lanosterol (1) is not a normal metabolite of higher plants and algae but that the first C-30 cyclized product in tetracyclic triterpene biosynthesis is cycloartenol (**2**). There is now a considerable literature which supports the view that cycloartenol probably replaces lanosterol as the key intermediate in plant sterol biosynthesis (see Goad and Goodwin, 1972, for full details). In particular it has been shown with the use of [2-^{14}C-(4*R*)-4-$^{3}H_1$] mevalonic acid (MVA) that cycloartenol is a primary product of cyclization of squalene 2,3-oxide and

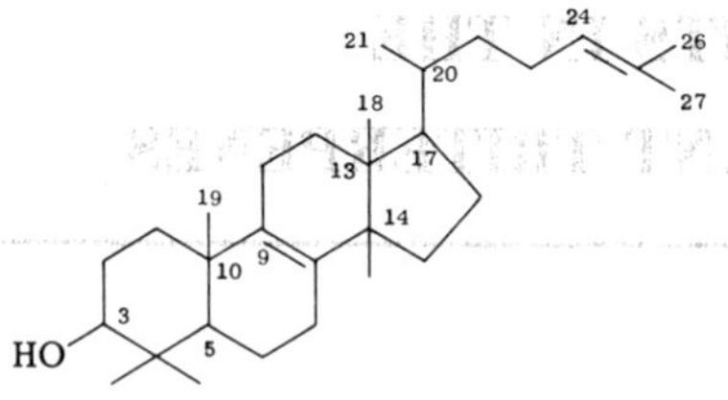

1 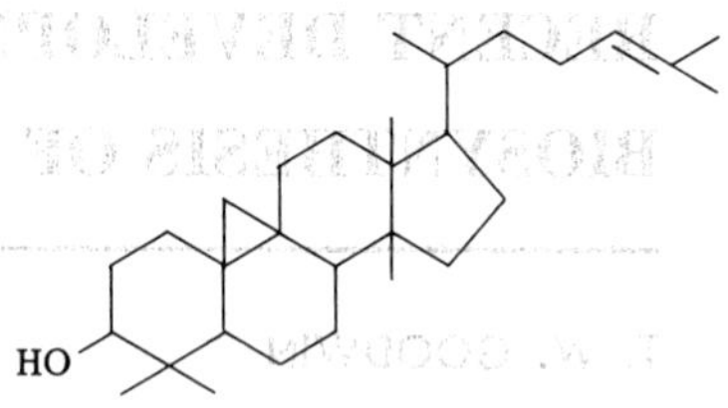 **2**

that it is not formed by isomerization of lanosterol (Rees *et al.*, 1968a; Goad and Goodwin, 1969); that no labeled lanosterol can be found when radioactive precursors are added to the plant material and lanosterol is present as a trap (Hewlins *et al.*, 1969); that under anaerobic conditions, a cell-free system from the alga *Ochromonas malhamensis* converts squalene 2,3-oxide only into cycloartenol (Rees *et al.*, 1968b, 1969). The enzyme has recently been partly purified, and its properties were compared with those

TABLE 1

THE PROPERTIES OF SQUALENE 2,3-OXIDE CYCLOARTENOL CYCLASE FROM *Ochromonas malhamensis*[a]

Substrate	Squalene 2,3-oxide
Product	Cycloartenol
Cell-Fraction	Microsomal
Effect of 0.15% (w/v) deoxycholate	2-Fold activation[b]
Optimum NaCl concentration	0.35 *M*
Effect of acetone-powdering	Enzyme activity much reduced
Approximate molecular weight	190,000

[a] See also Table 5.

[b] Maximum activation at 0.1 *M*.

TABLE 2

INCORPORATION OF LABELED CYCLOARTENOL, 24-METHYLENE CYCLOARTANOL AND LANOSTEROL INTO PORIFERASTEROL IN *Ochromonas malhamensis*[a,b]

	$[2\text{-}^3H_2]$Cycloartenol		$[2\text{-}^3H_2]$24-Methylene cycloartanol	$[2\text{-}^3H_2]$ Lanosterol
	Incubation 1	Incubation 2		
Radioactivity of precursor added (dpm)	2.20×10^6	6.40×10^6	2.20×10^6	3.30×10^6
In nonsaponifiable lipid (dpm)	7.63×10^5	—	1.15×10^6	—
In 4,4-dimethyl sterols[c] (dpm)	2.82×10^5	1.21×10^5	1.09×10^6	9.94×10^5
In 4α-methyl sterols[c] (dpm)	5.65×10^2	5.30×10^2	—	
In 4-demethyl sterols[c] (dpm)	1.96×10^5	4.10×10^5	1.18×10^5	1.04×10^5
In poriferasterol[d] (dpm)	9.89×10^4	2.78×10^5	1.02×10^5	9.32×10^4
Specific activity after addition of carrier poriferasterol (dpm/mg)				
Initial	2040	4610	—	—
First crystallization	1880	4500	6740	5640
Second crystallization	1770	4690	6540	5560
Third crystallization	1970	4550	6400	5860
Poriferasterol tetrabromide (dpm/mg)				
First crystallization	—	2400	3400	2900
Second crystallization	—	2590	4000	3360
Third crystallization	—	2520	—	—
Percentage incorporation into poriferasterol[e]	4.5	4.3	4.6	2.8

[a] Lenton *et al.* (1971), by permission of Academic Press, New York.

[b] The radioactive 4,4-dimethyl sterols were distributed between three 72-hour-old cultures which were harvested after a further 43-hour incubation. The cultures were run concurrently except for incubation 2 with $[2\text{-}^3H_2]$cycloartenol.

[c] Obtained by thin-layer chromatography on silica gel.

[d] Obtained by thin-layer chromatography on $AgNO_3$-silica gel.

[e] Based upon the radioactivity recovered in the poriferasterol isolated by the on $AgNO_3$-silica gel.

of squalene 2,3-oxide lanosterol cyclase (Table 1) (Beastall *et al.*, 1971). Labeled cycloartenol is effectively converted into phytosterols in *O. malhamensis* (Hall *et al.*, 1969; Lenton *et al.*, 1971) and in tissue cultures of *Nicotiana tabacum* (Hewlins *et al.*, 1969). On the other hand lanosterol is also an effective precursor of phytosterols in *Euphorbia peplus* (Baisted *et al.*, 1968), *O. malhamensis* (Hall *et al.*, 1969; Lenton *et al.*, 1971), *Zea mays* (Gibbons *et al.*, 1971), *N. tabacum* (Hewlins *et al.*, 1969), and *Pinus pinea* (Raabs *et al.*, 1968) (see, for example, Table 2). This is presumably a reflection of a relative lack of specificity of the enzymes concerned with the transformation of cycloartenol into phytosterols.

Routes from Cytoartenol to Phytosterols

Clearly many possibilities exist for the transformation of cycloartenol into phytosterols (4-demethyl sterols); for example, the existence of different mechanism of methylation at C-24 allows a number of alternatives (these are discussed later). If we confine ourselves to 24-methylene and 24-ethylidene derivatives as intermediates, then the pathway illustrated in Fig. 1 is a distinct possibility. All the proposed intermediates have been

TABLE 3

Incorporation of 24-Ethylidene Derivatives into Poriferasterol in *Ochromonas malhamensis*[a]

Substrate	Incubation time (hours)	Total activity added (dpm)	Activity re- in porifer- asterol (dpm)	Percent of initial activity in porifer- asterol
[23,25-$^3H_3$]Cycloeucalenol	110	1.34×10^7	2.41×10^6	18.0[a,b]
[2,4-$^3H_3$]Methylene lophenol	110	3.83×10^6	1.35×10^6	35.2
[2,4-$^3H_2$]Ethylidene lophenol	110	7.81×10^6	2.65×10^6	33.9
[2,2,44-$^3H_4$]5α-stigmasta-7,Z-24(28)-dien-3β-ol	144	2.79×10^7	2.18×10^6	7.8 [92][b,c]
[7,7-$^3H_2$]Fucosteryl acetate	144	2.42×10^7	9.02×10^5	3.7 [15][b,c]
[7,7-$^3H_2$]28-Isofucosteryl acetate	144	8.76×10^6	$1,38 \times 10^6$	15.7 (85][b,c]

[a] Lenton *et al.* (1971), by permission of Academic Press, New York, and Knapp *et al.* (1971).

[b] The tritium lost from C-23 introduction of Δ^{22} bond not considered in calculation.

[c] The percentage of total incorporation found in poriferasterol.

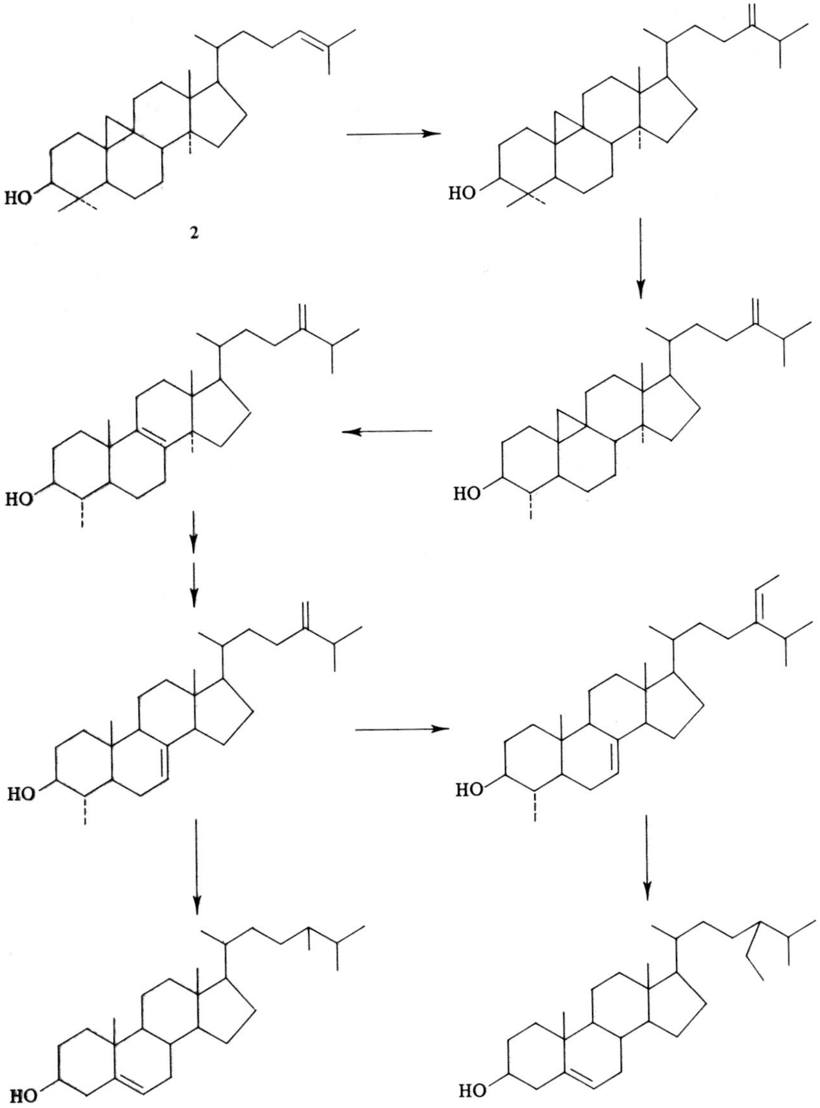

2

FIG. 1. A possible pathway for the formation of phytosterols from cycloartenol.

detected as natural products and, in particular, conversion of tritium-labeled cycloartenol, 24-methylene cycloartanol cycloeucalenol, 24-methylene lophenol, 24-ethylidene lophenol, stigmasta-7,Z-24(28)-dien-3β-ol, and 28-isofucosterol into poriferasterol has been established (Lenton *et al.*, 1971; Knapp *et al.*, 1971; Hall *et al.*, 1969) (Tables 2 and 3).

Removal of Methyl Groups at C-4 and C-14 in Phytosterol Biosynthesis

It would appear from the structures illustrated in Fig. 1 that one of the C-4 methyl groups is lost before the C-14 methyl group. This contrasts with cholesterol biosynthesis in animals where the C-14 methyl group is lost first (see Goad and Goodwin, 1972). The presence of the 9β-19-cyclopropane ring in cycloartenol and early intermediates hinders the methyl group which may prevent its attack by the appropriate demethylase; furthermore C-14 decarboxylation is facilitated by a double bond at C-7 or C-8 (see Goad and Goodwin, 1972) which can only exist after the opening of the cyclopropane ring. The point at which demethylation at C-14 occurs would appear to be at the 24-methylene- or 24-ethylidene-lophenol stage (see Fig. 1).

The order of removal of the methyl groups at C-4 was investigated in the fern *Polypodium vulgarie.* By the use of the degradation outlined in Fig. 2 and [2-^{14}C-4R-^{3}H$_1$] MVA, it was shown that the 4α-methyl group of cycloartenol was derived from C-2 of MVA and that it is this group which is lost when cycloartenol is converted into 31-norcycloartenol (**3**) (Ghisalberti *et al.,* 1969a). Thus it follows that the 4β-methyl group of cycloartenol takes up the 4α- position in 31-norcycloartenol. The same situation exists in the conversion of 24-methylene cycloartanol into cycloeucalenol

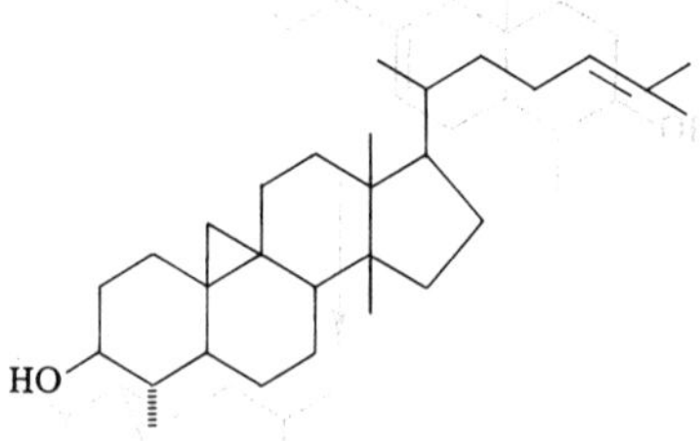

HO

3

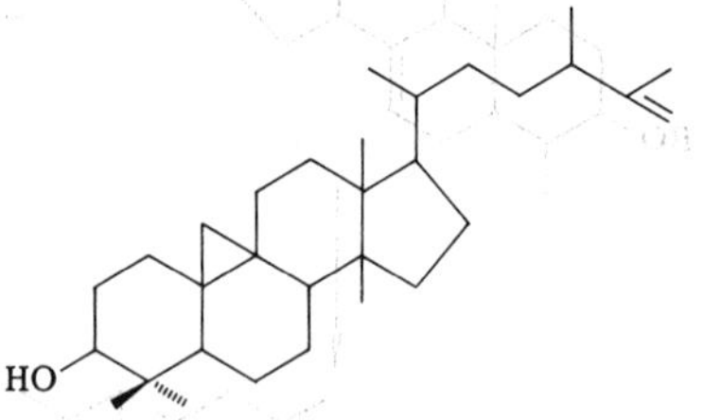

HO

4

HO

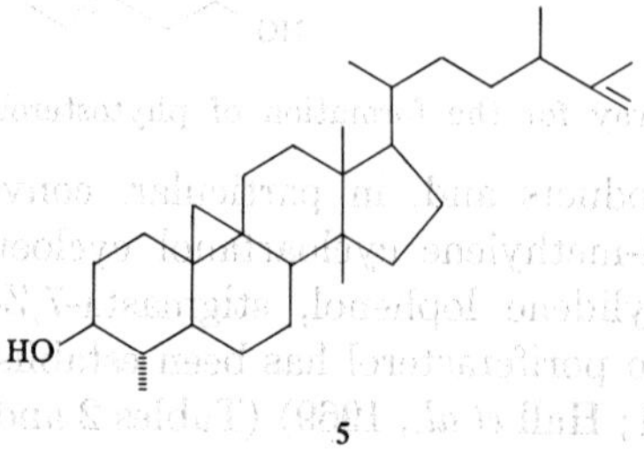

5

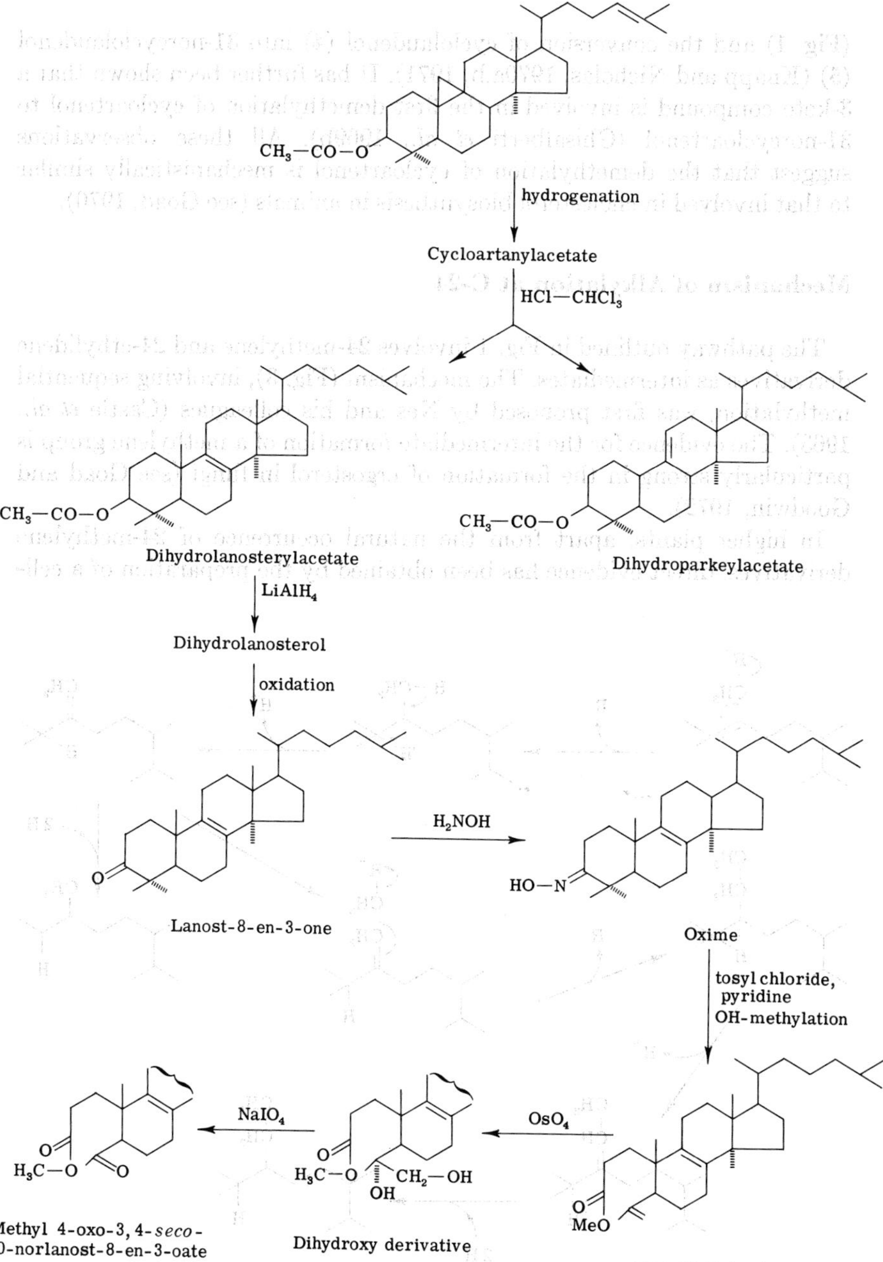

FIG. 2. Degradation of cycloartenyl acetate to identify the 4α- and 4β-methyl groups (Ghisalberti *et al.*, 1969b).

(Fig. 1) and the conversion of cyclolaudenol (4) into 31-norcyclolaudenol (5) (Knapp and Nicholas, 1970a,b, 1971). It has further been shown that a 3-keto compound is involved in the first demethylation of cycloartenol to 31-norcycloartenol (Ghisalberti *et al.*, 1969b). All these observations suggest that the demethylation of cycloartenol is mechanistically similar to that involved in cholesterol biosynthesis in animals (see Goad, 1970).

Mechanism of Alkylation at C-24

The pathway outlined in Fig. 1 involves 24-methylene and 24-ethylidene derivatives as intermediates. The mechanism (Fig. 3), involving sequential methylation, was first proposed by Nes and his colleagues (Castle *et al.*, 1963). The evidence for the intermediate formation of a methylene group is particularly strong in the formation of ergosterol in fungi (see Goad and Goodwin, 1972).

In higher plants, apart from the natural occurrence of 24-methylene derivatives, direct evidence has been obtained by the preparation of a cell-

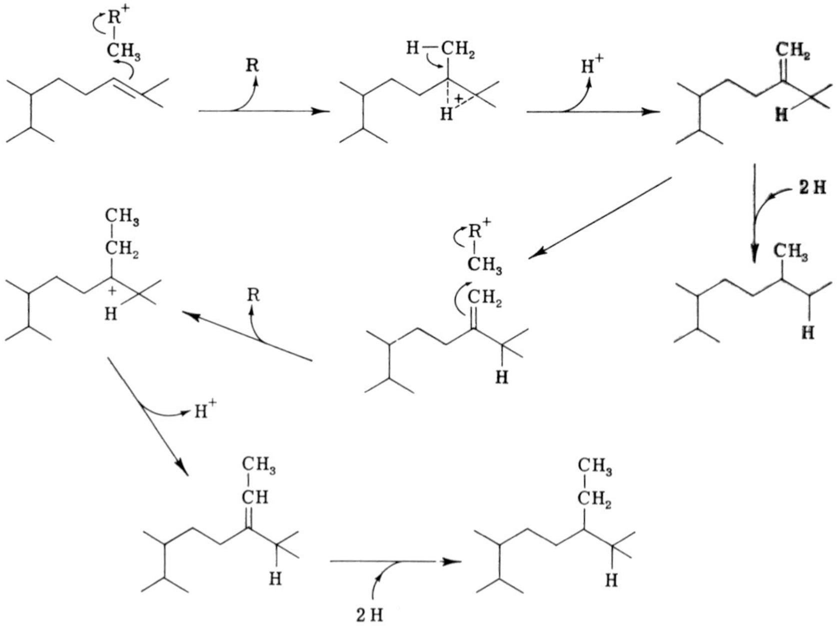

R^+ CH_3 R H—CH_2 H H^+ CH_2 H 2 H CH_3 H CH_3 CH_2 H R R^+ CH_3 CH_2 H H^+ CH_3 CH H 2 H CH_3 CH_2 H

FIG. 3. Mechanism of alkylation at C-24 involving 24-ethylidene intermediates (Castle *et al.*, 1967).

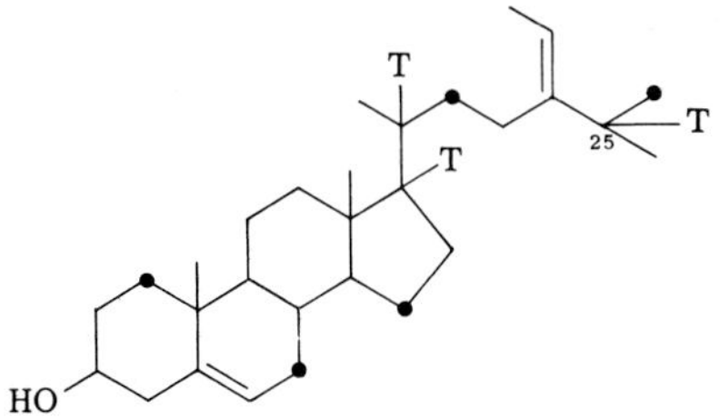

FIG. 4. Labeling of fucosterol from [2-^{14}C-(4R)-4-^{3}H$_1$]MVA (Goad and Goodwin, 1969). (● = C-2 of MVA; T = 4R-4^{3}H$_1$ of MVA)

free system from pea seedlings which converts cycloartenol into 24-methylene cycloartanol (Castle *et al.*, 1967), and by the demonstration of hydrogen migration from C-24 to C-25 in the formation of the latter in *Fucus spiralis* (Goad and Goodwin, 1969) and in *Nicotiana tabacum* (Tomita and Uomori, 1970).

Introduction of the second methyl group as indicated in Fig. 3 demands a hydrogen migration from C-24 to C-25, the retention of four not five hydrogens from the two methyl groups of methionine, and the demonstration that an ethylidene derivative can be reduced to an ethyl derivative. The first requirement has been met as indicated by the results of studies with [2-^{14}C-(4R)-4-^{3}H$_1$]MVA; this substrate labels cycloartenol with tritium at C-17, C-20, and C-24. Fucosterol synthesized from the same substrate also contained three tritium atoms, but in this case one of them could not be located at C-24. Degradations showed that it was located at C-25 (Fig. 4) (Goad and Goodwin, 1965; Goad *et al.*, 1966). Similar results were obtained with poriferasterol (**6**) synthesized by *O. malhamensis* (Smith, 1969). Experiments with pine seedlings using [24-^{3}H$_1$] lanosterol also yielded 28-isofucosterol (**7**) with a tritium at C-25 (Raabs *et al.*, 1968).

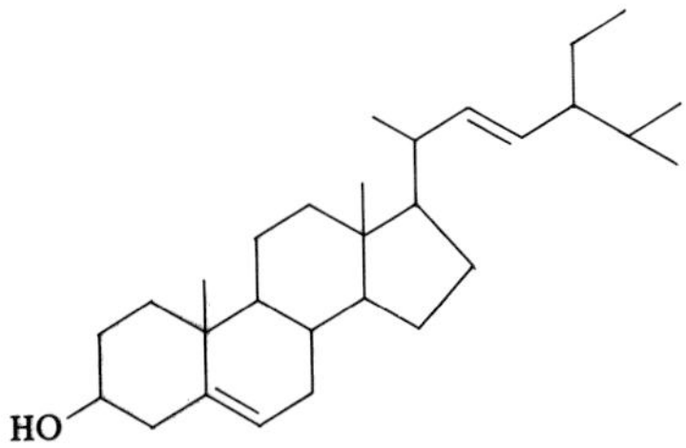

6

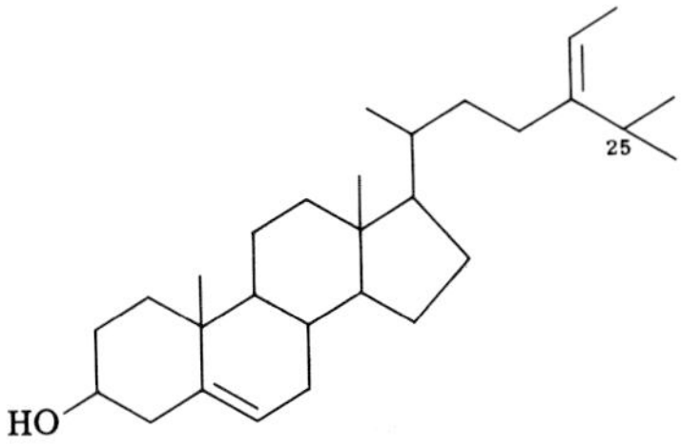

7

The second criterion was established in *Ochromonas malhamensis* and *O. danica* using [C^{2}H$_3$]methionine. The isolated poriferasterol contained species with 1, 3, and 4 deuterium atoms located in the C-24 ethyl group but no indication of a species with five deuterium atoms (Smith *et al.*, 1967; Lenfant, unpublished, quoted by Lederer, 1969). This pattern had previously been indicated using [C^3H_3]methionine, but the result was not unequivocal because of the possibility of a large tritium isotope effect (Goad *et al.*, 1966). As already indicated the utilization of 24-ethylidene sterols as precursors of 24-ethyl sterols has been demonstrated for a number of compounds (Table 3) (Lenton *et al.*, 1971; Knapp *et al.*, 1971). It will be noted that 28-isofucosterol is very much more efficiently incorporated than its epimer fucosterol, which requires explanation. It is likely that the 28-isofucosterol configuration is that of the *in vivo* precursor, and the reductase involved would show some specificity. It is also possible that in stigmasterol and poriferasterol biosynthesis a Δ^{22} bond is introduced before the reduction of the $\Delta^{24(28)}$ bond and the fucosterol configuration might hinder the desaturation at C-22. Observations with the protozoan *Tetrahymena pyriformis* would support this view; it will metabolize 28-isofucosterol to stigmasta-5,7,22-Z-24(28)-tetraen-3β-ol, but fucosterol is converted into the 5,7,E-24(28)-trien-3β-ol; that is, insertion of the Δ^{22}

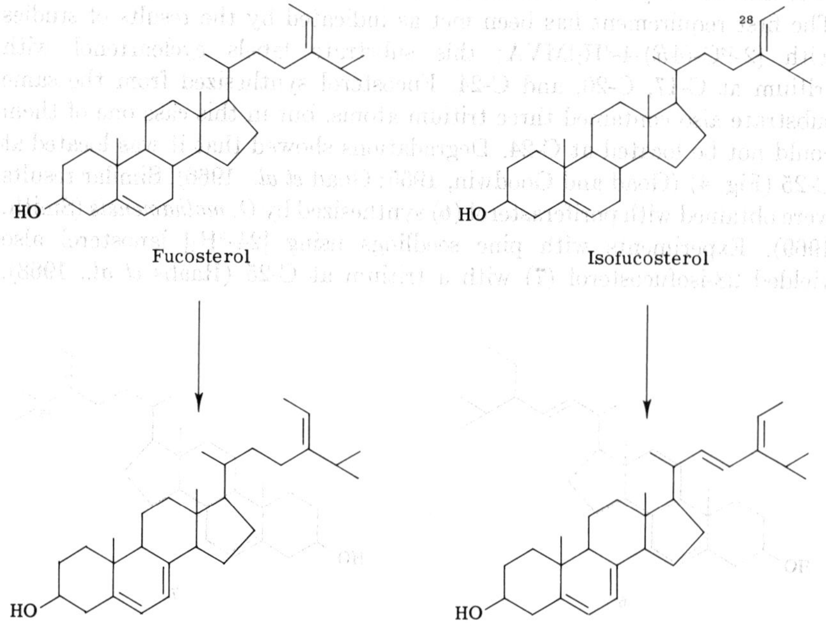

Fig. 5. Metabolism of fucosterol and 28-isofucosterol by *Tetrahymena pyriformis* (Nes *et al.*, 1971).

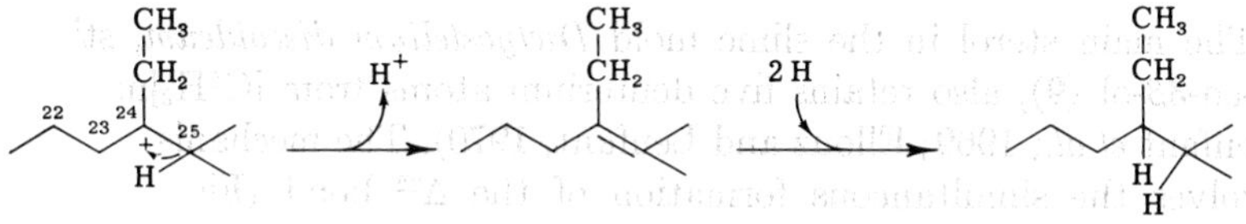

Fig. 6. Mechanism of alkylation at C-24 involving $\Delta^{2(25)}$ intermediates (Tomita *et al.*, 1970).

double bond occurs only with 28-isofucosterol (Fig. 5) (Nes, 1970; Nes *et al.*, 1971).

Other alkylation mechanisms must also exist which do not involve 24-ethylidene intermediates. Chondrillasterol (**8**) and 22-dihydrochondrillasterol synthesized by *Chlorella vulgaris* in the presence of $[C^2H_3]$-methionine contain five deuterium atoms (Tomita *et al.*, 1970), as does poriferasterol in *C. ellipsoidea* (Tomita *et al.*, 1971). The mechanism pro-

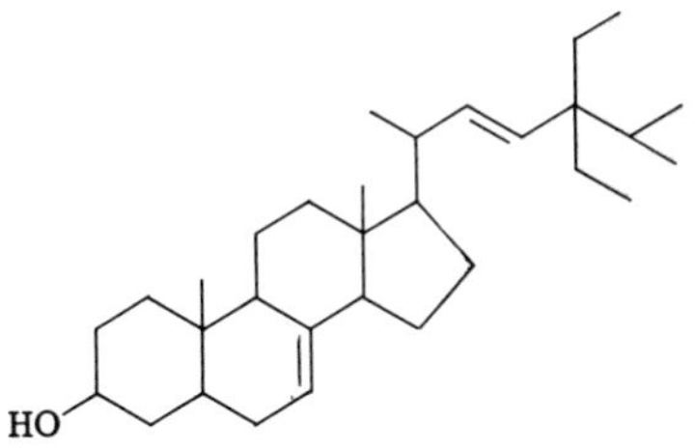

8

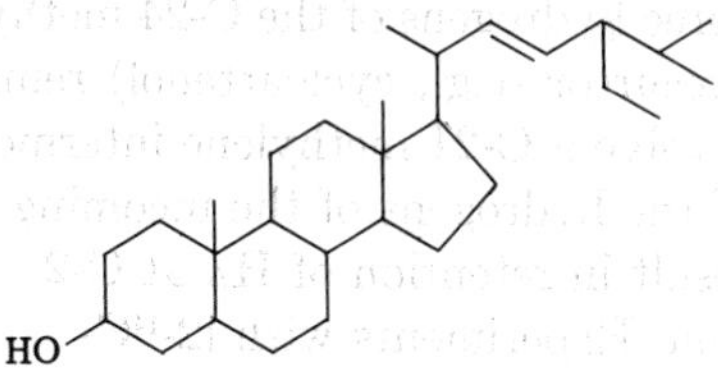

9

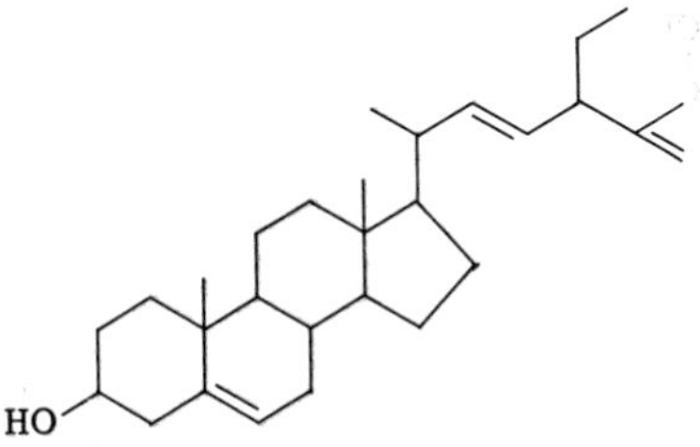

10

posed is illustrated in Fig. 6. As yet there is no evidence for the natural occurrence of a sterol with a $\Delta^{24(25)}$ double bond, but support for the mechanism comes from the observation that stigmasterol formed by tissue cultures of *Nicotiana tabacum* and *Dioscorea tokoro* in the presence of $[(4R)\text{-}4\text{-}^3H_1]$MVA retains only 2 tritium atoms neither of which is located at C-24 or C-25 (Tomita and Uomori, 1970).

The main sterol in the slime mold *Dictyostelium discoideum*, stigmast-22-en-3β-ol (**9**), also retains five deuterium atoms from [C^2H_3]methionine (Lenfant *et al.*, 1969; Ellouz and Lenfant, 1970). The mechanism proposed involves the simultaneous formation of the Δ^{22} bond (Fig. 7). This is supported by the observation that [23-$^3H_2$]lanosterol is incorporated into stigmast-22-en-3β-ol in *D. discoideum* with retention of tritium at both C-23 and C-24 (Ellouz and Lenfant, 1969a, 1970). However, in the same organism stigmastan-3β-ol can be converted in stigmast-22-en-3β-ol, which means that C-24 alkylation can also occur independently of the formation of the Δ^{22} double bond (Ellouz and Lenfant, 1969b).

Another alkylation mechanism is linked with the formation of a Δ^{25-26} bond, as in the formation of cyclolaudenol (**4**) and 24*S*-24-ethylcholesta-5,22,25-trien-3β-ol (**10**). Three possibilities exist for the methylation and Δ^{25-26} double bond formation in cyclolaudenol (Fig. 8). Route (a) involves elimination of a proton from C-26 from the intermediate cation indicated, which would give the cyclolaudenol side chain directly with retention of all three hydrogens of the C-24 methyl group and with the hydrogen (H_A) of a precursor (e.g., cycloartenol) remaining at C-24. Routes (b) and (c) both involve a C-24 methylene intermediate and thus the retention of only two of the hydrogens of the incoming methyl group; however, route (c) would result in retention of H_A at C-24 and route (b) would result in its elimination. Experiments with [2-^{14}C-(4*R*)-4-$^3H_1$]MVA showed that cyclolaudenol formed by rhizomes of *Polypodium vulgare* retained H_A at C-24 and that the methylene group at C-25 arose stereospecifically from C-2 of MVA. Thus

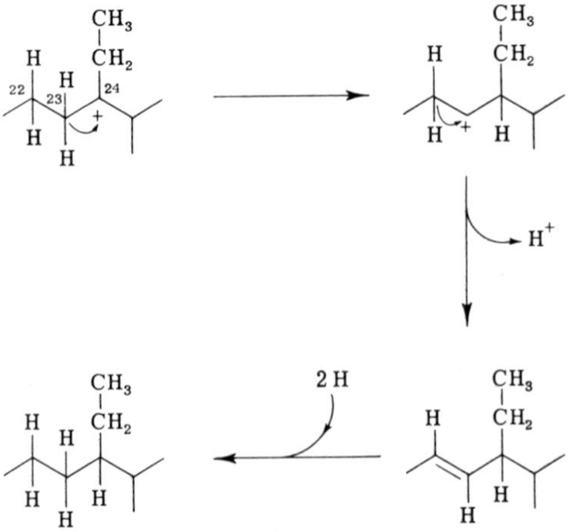

FIG. 7. Mechanism of alkylation at C-24 involving Δ^{22} intermediates (Lederer, 1969).

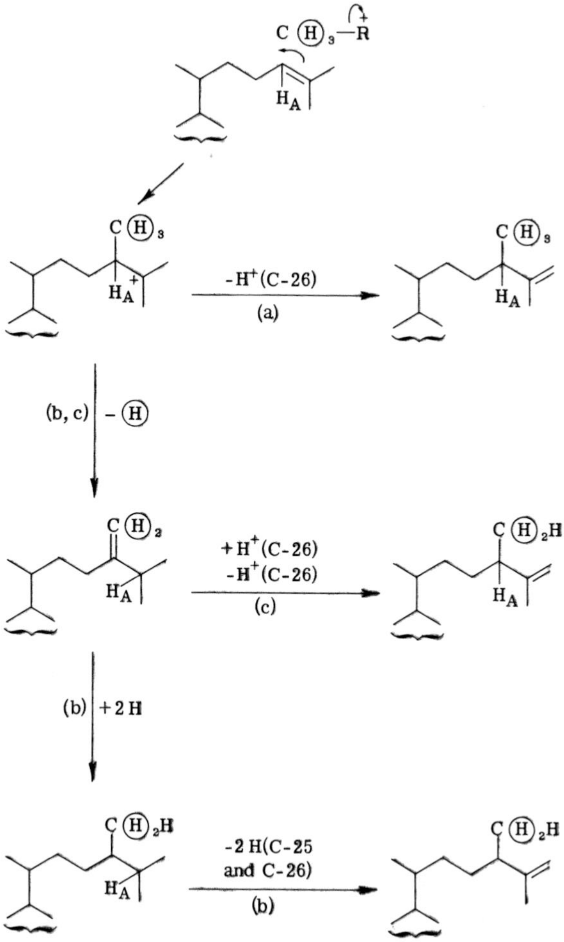

FIG. 8. Possible mechanism of methylation to produce cyclolaudenol (Ghisalberti *et al.*, 1969b). The circled H is derived from the methyl group of methionine, and H_A arises from the 4-pro-*R* hydrogen of mevalonic acid.

route (c) was eliminated; evidence which favors route (a) was obtained by comparing the ^3H:^{14}C ratio in cyclolaudenol and 24-methylene cycloartanol synthesized from [^{14}C:$^3H_3$]methionine by *P. vulgare* in the same experiment. The ratio in cyclolaudenol (17·4:1) was very close to that of the starting methionine (16·5:1) whereas that in 24-methylene cycloartanol was significantly lower (14·1:1) (Ghisalberti *et al.*, 1969b). This result needs confirmation by use of [C^2H_3]methionine, in order to eliminate any possible isotope effect.

TABLE 4

INCORPORATION OF [2-^{14}C-(4*R*)-4-$^{3}H_1$]MEVALONATE INTO 24-ETHYLCHOLESTA-5,22,25-TRIEN-3β-OL IN *Clerodendrum campbellii*[a]

Compound	$^{3}H:^{14}C$ radioactivity ratio	$^{3}H:^{14}C$ atomic ratio, based on squalene
[2-^{14}C-(4*R*)-4-$^{3}H_1$]Mevalonate	6.43:1.0	—
Squalene	6.33:1.0	—
(24*S*)-Ethylcholesta-5,22,25-trien-3β-yl acetate	3.73:1.0	2.95:5
(24*S*)-Ethylcholesta-5,22-dien-3β,25ξ,26-triol 3-acetate	3.78:1.0	2.99:5
(24*S*)-24-Ethyl-25-oxo-26-norcholesta-5,22-dien-3β-yl acetate	3.72:1.0	2.96:5
24-Ethyl-25-oxo-26-norcholesta-5,23-dien-3-yl acetate	2.60:1.0	2.05:5
24-Methylenecycloartanyl acetate	6.30:1.0	5.98:6

[a] Bolger *et al.*, 1970b. Reprinted from the *Biochemical Journal*, by permission.

In the case of 24*S*-24-ethylcholesta-5,22,25-trien-3β-ol, first isolated from *Clerodendrum campbellii* (Bolger *et al.*, 1970a), experiments with [2-^{14}C-(4*R*)-4-$^{3}H_1$]MVA showed that the tritium originally at C-24 in a non-alkylated precursor, was still present in the same position in the triene (Table 4) (Bolger *et al.*, 1970b).

So the mechanism indicated in Fig. 9 is probably operating in producing a simultaneous alkylation and desaturation at C-24.

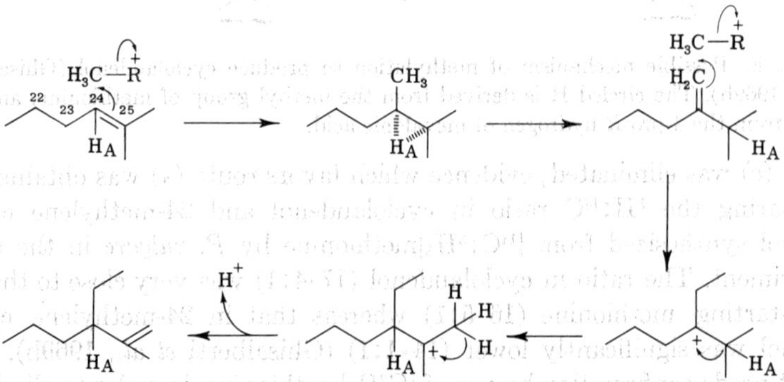

FIG. 9. Mechanism of alkylation to produce 24-ethyl-Δ^{25} sterols (Bolger *et al.*, 1970b).

Pentacyclic Triterpene Hydrocarbons

The pentacyclic hydrocarbons, diploptene (**11**), serratene (**12**), hopene I, and fernene are formed in *Polypodium vulgare* by proton-catalyzed cyclization of squalene and not via squalene 2,3-oxide (Barton *et al.*, 1969). Using [2-^{14}C-(4*R*)-4-$^3H_1$]MVA it was possible to show that all 6 tritium atoms arising from C-4 of MVA in squalene are retained in fernene and only five in hopene I (Ghisalberti *et al.*, 1970).

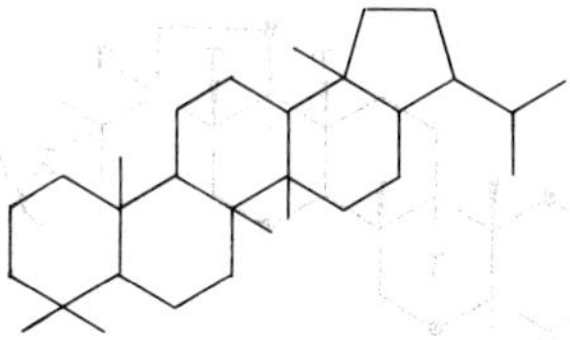

11

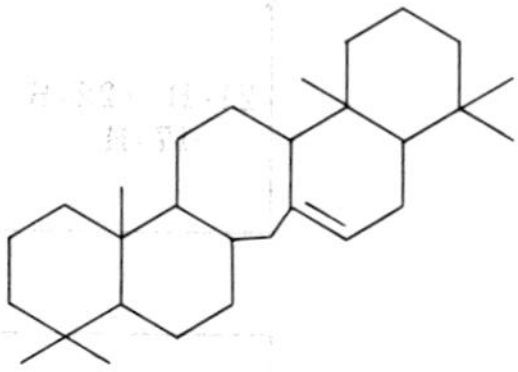

12

Thus no double bonds are involved in the final rearrangement leading to fernene. A possible mechanism starting with squalene in the chair-chair-chair-chair-boat conformation is indicated in Fig. 10. Seven 1,2 shifts are

TABLE 5

A COMPARISON OF THE CYCLASES FROM PIG LIVER, YEAST, AND *Ononis spinosa*

Source	Liver[a]	Yeast[a]	*Ononis spinosa*[b]
Product	Lanosterol	Lanosterol	α-Onocerin
Activity (nmoles/mg/hr)	4–47	1	0.2
Cell fraction	Microsomal	Soluble	Soluble
Effect of 0.15% (w/v) deoxycholate	3-Fold activation	None	80% Inhibition
Effect of 0.2% (w/v) Triton X-100	None	6-Fold activation	85% Inhibition
Optimum K^+ concentration (*M*)	0.4	ca 0.01	0.15
Effect of acetone powdering	Enzyme activity lost, but restored by deoxycholate	Enzyme retains activity but is precipitated	Enzyme remains soluble and active

[a] Data from Schechter *et al.* (1970).
[b] Data from Rowan and Dean (1972).

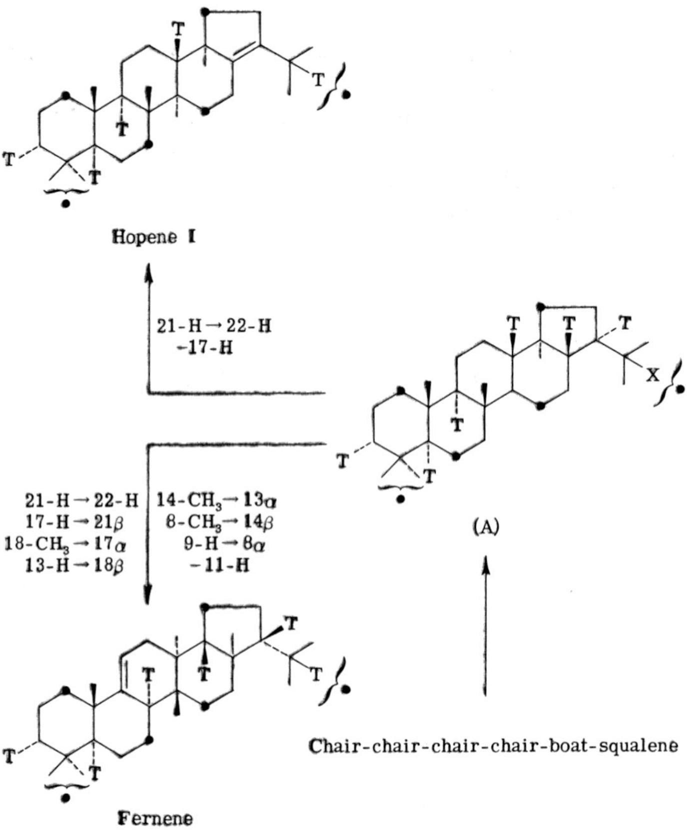

Hopene I

21-H → 22-H
-17-H

(A)

21-H → 22-H
17-H → 21β
18-CH_3 → 17α
13-H → 18β

14-CH_3 → 13α
8-CH_3 → 14β
9-H → 8α
-11-H

Chair-chair-chair-chair-boat-squalene

Fernene

FIG. 10. Mechanism for cyclization of squalene to produce fernene and hopene I (Ghisalberti *et al.*, 1970).

proposed in which all 6 tritiums arising from C-4 of MVA are retained. A possible pathway to hopene I is also indicated.

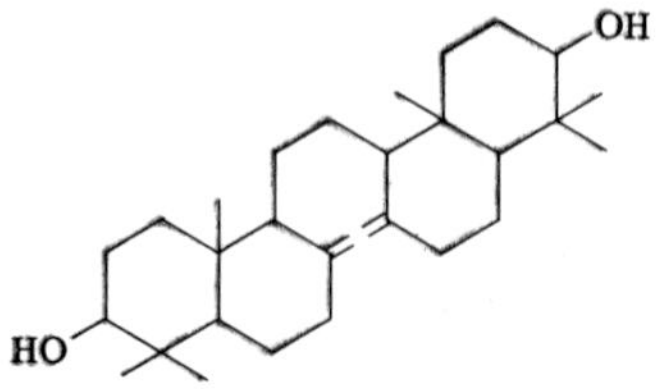

OH

HO

13

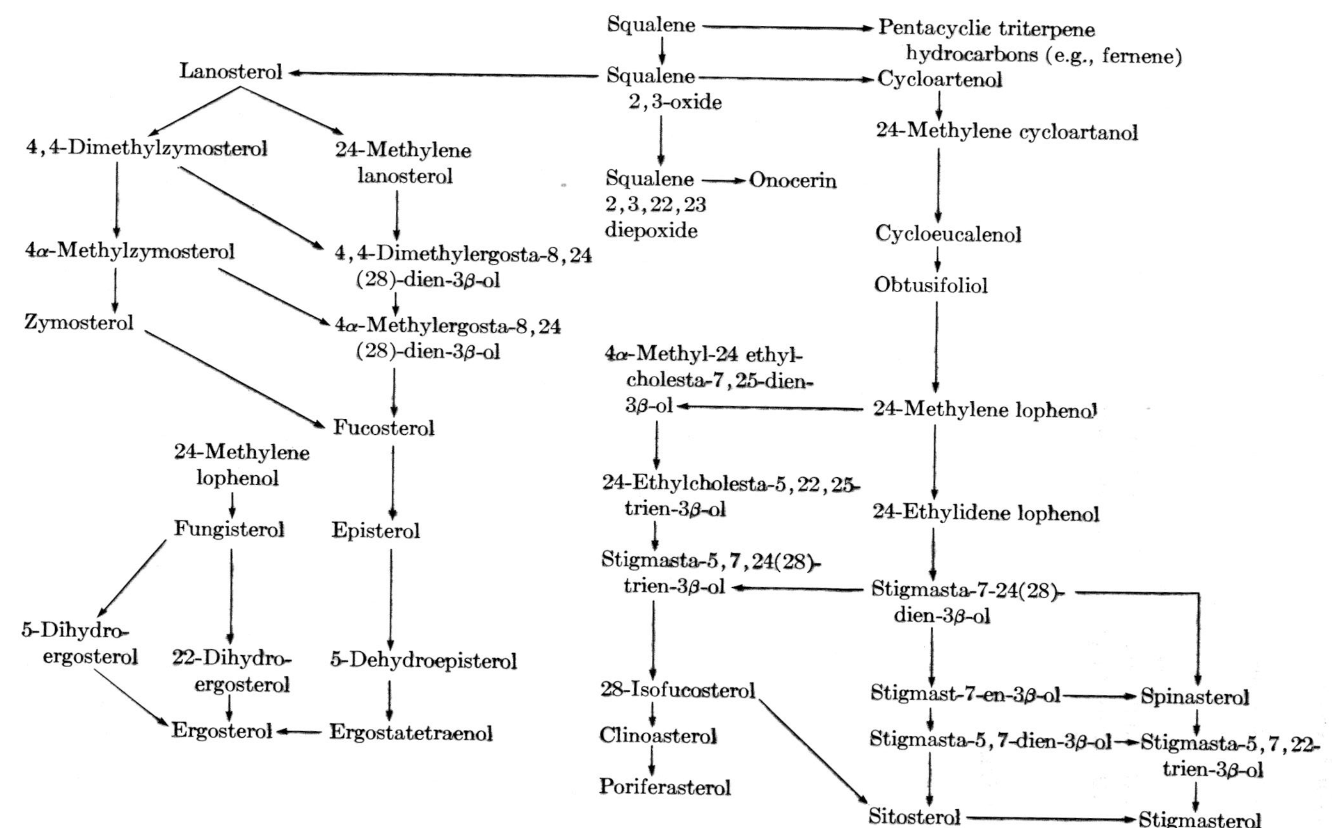

Fig. 11. General pathway of triterpene biosynthesis in higher plants and algae (cycloartenol pathway) and in fungi (lanosterol pathway).

Onocerin Biosynthesis

A possible intermediate in the production of onocerin (**13**) by *Ononis* spp. is squalene 2,3;22,23-diepoxide, with cyclization proceeding from both ends of the molecule. This idea has been experimentally demonstrated in a cell-free preparation from *Ononis spinosa* (Rowan and Dean, 1972). The properties of the squalene, 2,3;22,23-diepoxide-onocerin cyclase have been compared with those of squalene, 2,3-oxide-lanosterol cyclase in yeast and liver (Table 5).

Conclusions

The work described here taken in conjunction with many other observations from numerous laboratories (see, e.g., Goodwin, 1971; Goad and Goodwin, 1972) allows reasonably clear biosynthetic pathways to be drawn for triterpene synthesis.

In the case of the plant sterols, it is clear that in higher plants and algae the first intermediate is cycloartenol whereas in fungi it is lanosterol (Fig. 11). The subsequent steps, most of which have been experimentally demonstrated, represent an interdigitating complex, for which it is very difficult to decide which, if any, is the major pathway. Indeed different pathways resulting in the same final product may predominate in different organisms. The elucidation of this aspect of plant sterol biosynthesis is very much for the future and its success will depend to a great extent on the development techniques to deal enzymatically with highly hydrophobic substrates.

Acknowledgments

The work from the author's laboratory has been supported by the Science Research Council. I am most indebted to Dr. L. J. Goad and Dr. H. H. Rees for their expert collaboration. The achievements of postdoctoral fellows and graduate students are also gratefully acknowledged.

References

Baisted, D. J., R. L. Gardner, and L. A. McReynolds. 1968. *Phytochemistry* **7**:945.

Barton, D. H. R., A. F. Gosden, G. Mellows, and D. A. Widdowson. 1969. *Chem. Commun.* p. 184.

Beastall, G. H., H. H. Rees, and T. W. Goodwin. 1971. *FEBS* (*Fed. Eur. Biochem. Soc.*), *Lett.*

Bolger, L. M., H. H. Rees, E. L. Ghisalberti, L. J. Goad, and T. W. Goodwin. 1970a. *Tetrahedron Lett.* No. 35, p. 3043.

Bolger, L. M., H. H. Rees, E. L. Ghisalberti, L. J. Goad, and T. W. Goodwin. 1970b. *Biochem. J.* **118**:197.

Castle, M., G. Blondin, and W. R. Nes. 1963. *J. Amer. Chem. Soc.* **85**:3306.

Castle, M., G. A. Blondin, and W. R. Nes. 1967. *J. Biol. Chem.* **242**:5796.

Ellouz, R., and M. Lenfant. 1969a. *Tetrahedron Lett.* p. 609.

Ellouz, R., and M. Lenfant. 1969b. *Tetrahedron Lett.* p. 2655.

Ellouz, R., and M. Lenfant. 1970. *Tetrahedron Lett.* p. 3967.

Ghisalberti, E. L., N. J. de Souza, H. H. Rees, L. J. Goad, and T. W. Goodwin. 1969a. *Chem. Commun.* p. 1403.

Ghisalberti, E. L., N. J. de Souza, H. H. Rees, L. J. Goad, and T. W. Goodwin. 1969b. *Chem. Commun.* p. 1401.

Ghisalberti, E. L., N. J. de Souza, H. H. Rees, and T. W. Goodwin. 1970. *Phytochemistry* **9**:1817.

Gibbons, G. F., L. J. Goad, T. W. Goodwin, and W. R. Nes. 1971. *J. Biol. Chem.* **246**: 3967.

Goad, L. J. 1970. *Biochem. Soc. Symp.* **29**:45.

Goad, L. J., and T. W. Goodwin. 1965. *Biochem. J.* **96**: 79P.

Goad, L. J., and T. W. Goodwin. 1969. *Eur. J. Biochem.* **7**:502.

Goad, L. J., and T. W. Goodwin. 1972. *In* "Progress in Phytochemistry," p. 113. Wiley (Interscience), New York.

Goad, L. J., A. S. A. Hamman, A. Dennis, and T. W. Goodwin. 1966. *Nature* (*London*) **210**:1322.

Goodwin, T. W. 1971. *Biochem. J.* **123**:293.

Hall, J., A. R. H. Smith, L. J. Goad, and T. W. Goodwin. 1969. *Biochem. J.* **112**:129.

Hewlins, M. J. E., J. D. Ehrhardt, L. Hirth, and G. Ourisson. 1969. *Eur. J. Biochem.* **8**:184.

Knapp, F. F., and H. J. Nicholas. 1970a. *Steriods* **16**:329.

Knapp, F. F., and H. J. Nicholas. 1970b. *Chem. Commun.* p. 399.

Knapp, F. F., and H. J. Nicholas, 1971. *Phytochemistry* **10**:97.

Knapp, F. F., J. B. Greig, L. J. Goad, and T. W. Goodwin. 1971. *J. Chem. Soc. D* p. 707.

Lederer, E. 1969. *Quart Rev. Chem. Soc.* **23**:453.

Lenfant, M., R. Ellouz, B. C. Das, E. Lissman, and E. Lederer. 1969. *Eur. J. Biochem.* **7**:159.

Lenton, J. R., J. Hall, A. R. H. Smith, E. L. Ghisalberti, H. H. Rees, L. J. Goad, and T. W. Goodwin. 1971. *Arch. Biochem. Biophys.* **143**:664.

Nes, W. R. 1970. *J. Amer. Oil Chem. Soc.* **47**:85A.

Nes, W. R., P. A. A. Malya, F. B. Mallory, K. A. Ferguson, J. R. Landrey, and R. C. Conner. 1971. *J. Biol. Chem.* **246**:561.

Raabs, K. H., N. J. de Souza, and W. R. Nes. 1968. *Biochim. Biophys. Acta* **152**:742.

Rees, H. H., L. J. Goad, and T. W. Goodwin. 1968a. *Biochem. J.* **107**:417.

Rees, H. H., L. J. Goad, and T. W. Goodwin. 1968b. *Tetrahedron Lett.* No. 6, p. 723.

Rees, H. H., L. J. Goad, and T. W. Goodwin. 1969. *Biochim. Biophys. Acta* **176**:892.

Rowan, M., and P. G. D. Dean. 1972. *Phytochemistry* **11**:3111.

Schechter, I., F. W. Sweat, and K. Bloch. 1970. *Biochim. Biophys. Acta* **220**:463.

Smith, A. R. H. 1969. Ph.D. Thesis, University of Liverpool, Liverpool.

Smith, A. R. H., L. J. Goad, T. W. Goodwin, and E. Lederer. 1967. *Biochem. J.* **104**:56C.

Tomita, Y., and A. Uomori. 1970. *Chem. Commun.* p. 1416.

Tomita, Y., A. Uomori, and H. Minato. 1970. *Phytochemistry* **9**:555.

Tomita, Y., A. Uomori, and E. Sakurai. 1971. *Phytochemistry* **10**:573.

MECHANISMS OF INDOLE ALKALOID BIOSYNTHESIS. RECOGNITION OF INTERMEDIACY AND SEQUENCE BY SHORT-TERM INCUBATION

A. I. SCOTT, P. B. REICHARDT,* M. B. SLAYTOR,† and J. G. SWEENY‡

Sterling Chemistry Laboratory, Yale University, New Haven, Connecticut

Introduction

The state of the art of higher plant biosynthesis can fairly be said to be entering a new phase of development. Progressing from speculations

* Present address: Department of Chemistry, University of Alaska, Fairbanks, Alaska.

† Present address: Department of Biochemistry, The University of Sydney, Sydney NSW 2006, Australia.

‡ Present address: Organic Chemistry Laboratories, University of Technology, Loughborough, Leicestershire, England.

throughout the first half of the century, pioneered in large measure by Sir Robert Robinson (1917, 1955), the advent of tracer methods and the establishment of criteria for biointermediacy during the last two decades has led to an understanding of many of the pathways by which complex natural products are formed in whole plant systems. However, it is clear, especially in the alkaloid series, that the demonstration of specific incorporation without randomization of label together with the proof for the presence of a postulated intermediate are frequently difficult to obtain with plant material (Spenser, 1968). In particular, it is felt that low specific incorporations (0.001–0.01 percent), although satisfying acceptable conventions of constant radioactivity and in many (but not all) cases supported be degradative and/or multiple labeling data, may not distinguish between obligatory biointermediacy and nonspecific biotransformation. Problems such as nonpermeability of complex intermediates and large, variable pool sizes in mature biological material are often more serious in plant systems than in mammalian and fungal metabolism, where so much is known of controlled culture conditions.

Thus the third phase of the study of plant biosynthesis is now under way, and methods are being sought for the elucidation of pathways at the cell-free level. Mention may be made of recent terpenoid studies (Banthorpe and Wirz-Justice, 1969; Battaile *et al.*, 1968; Schechter and West, 1969), a study of N- and O-methylation (Mudd, 1960; Fales *et al.*, 1963), and earlier experiments with $^{14}CO_2$ (Rapoport *et al.*, 1960). However, by comparison with our knowledge of the path of carbon in photosynthesis (e.g., see Walker and Crofts, 1970), the subject of short-term analysis of the biosynthesis of alkaloids and other complex plant products is virtually unexplored.

Thus it became of interest to devise methods for distinguishing between the static and dynamic constituents of a plant such as *Vinca rosea* which at maturity contains as many as 100 alkaloids yet whose seeds are virtually devoid of alkaloidal material (Scott, 1970). Some progress in obtaining a cell-free system using acetone powders from homogenates of 8- to 12-day-old seedlings of *V. rosea* has been made, but the activity of this preparation for alkaloid synthesis is still very low and so far the only reproducible partial reaction has been the decarboxylation of tryptophan to tryptamine (M. B. Slaytor, unpublished observations). During the course of these studies, however, it was found that specific incorporations of DL-tryptophan 2-^{14}C into the alkaloids of young seedlings was extremely high (ca. 30 percent). It was therefore decided to develop the technique of short-term incubation of tryptophan with young seedlings (9–17 days) grown from a mixed strain of *V. rosea* at 92°F using full illumination. Promising intermediates can, in principle, be detected by autoradiography of chromatograms corresponding to frequent sampling of the alkaloidal content. Provided that small pool sizes are present during experiments with "early" biological material and that a linear uptake of tryptophan is in operation during the times of assay, we could anticipate that pivotal intermediates should at appropriate points in the sequence contain a relatively large proportion of the total radioactivity, which would then decay as the counts are transferred to the next intermediate with a similar "radioprofile." At the same time the less dynamic constituents of *V. rosea* should be gradually gaining radioactivity as they are (irreversibly) laid down in labeled form.

The technique has the additional advantage that rapid evaluation can be made on a qualitative basis for several related species. Quantitative experiments require development of two-dimensional thin-layer chromatography (TLC) systems, removal and collection of the alkaloids and recrystallization (after dilution with authentic material) to constant radioactivity. For the isolation and identification of unknown metabolites with "dynamic" radioprofiles, it will probably be necessary to carry out large-scale cell-free incubations to accumulate sufficient material for structural determination. Refeeding experiments can be used to decide whether the unknown occupies a crucial position and in fact justifies further analysis.

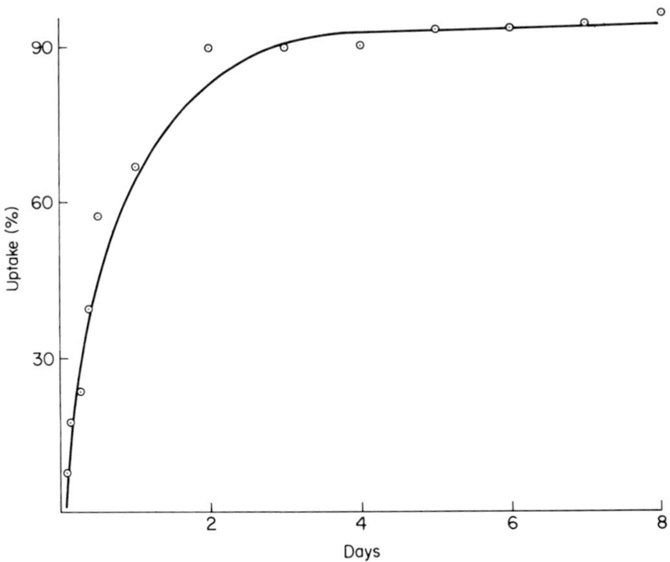

FIG. 1. Uptake of DL-tryptophan-^{14}C by *Vinca rosea* seedlings.

Although our experiments in this field are still incomplete, it has become evident that biointermediates can in fact be distinguished from other less dynamic constituents, and further that a 100-fold increase over previous specific incorporations can be demonstrated. In spite of the crude nature of the kinetic data which emerge from our experiments, we shall see that a number of conclusions regarding the early stages of alkaloid synthesis in *V. rosea* can be made.

Uptake of DL -Tryptophan-2-^{14}C

The incorporation of DL-tryptophan by the seedlings is shown in Fig. 1 where 0.06 μmole was fed to 4 seedlings for each indicated time interval. It can be seen that the maximum uptake (90 percent) is complete within 2–3 days. Reference to Fig. 2 reveals that quite large amounts (5 mg/g) can be absorged in a 10-hour experiment.

Incorporation of tryptophan into the alkaloids takes place in two distinct phases. For the first 2 hours of incubation it is linear (Figs. 3 and 4), but after this interval a rapid increase in the rate of incorporation is observed. Between 12 and 48 hours a maximum of 3 percent is reached and this level is maintained during the full time of the experiment (up to 8 days). It is

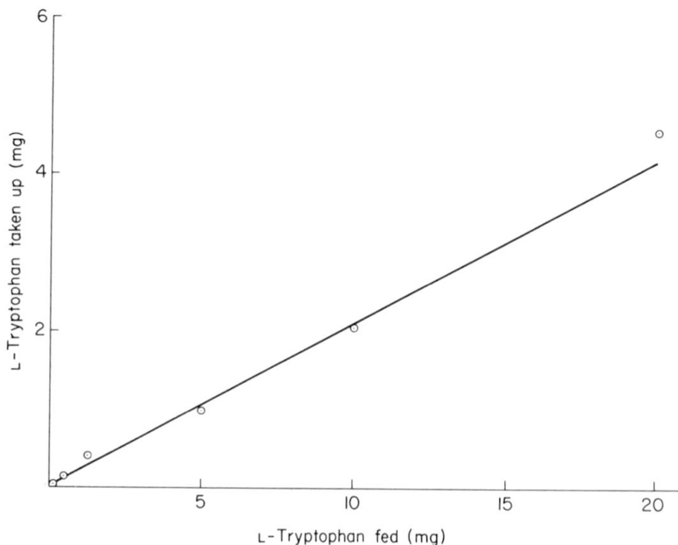

FIG. 2. Uptake of L-tryptophan by 1 g of seedlings.

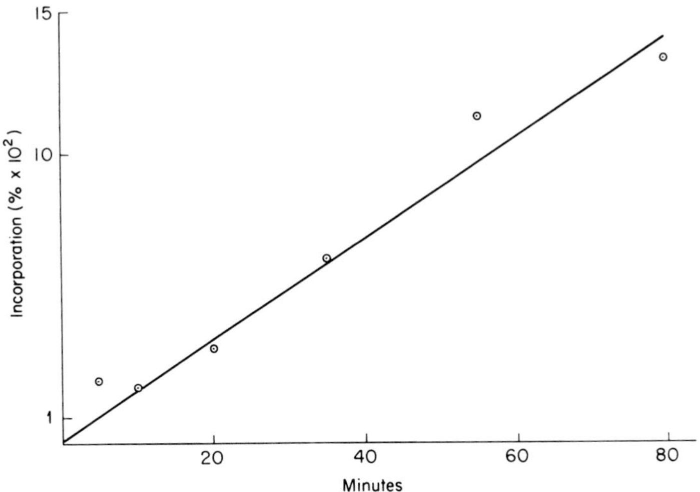

FIG. 3. Incorporation (percent) of DL-tryptophan-^{14}C into alkaloids in *Vinca rosea* seedlings during 80 minutes.

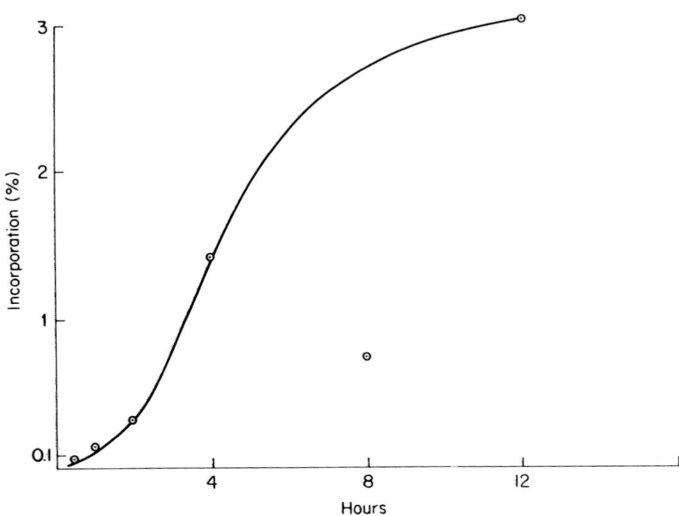

FIG. 4. Incorporation (percent) of DL-tryptophan-^{14}C into alkaloids in *Vinca rosea* seedlings during 12 hours.

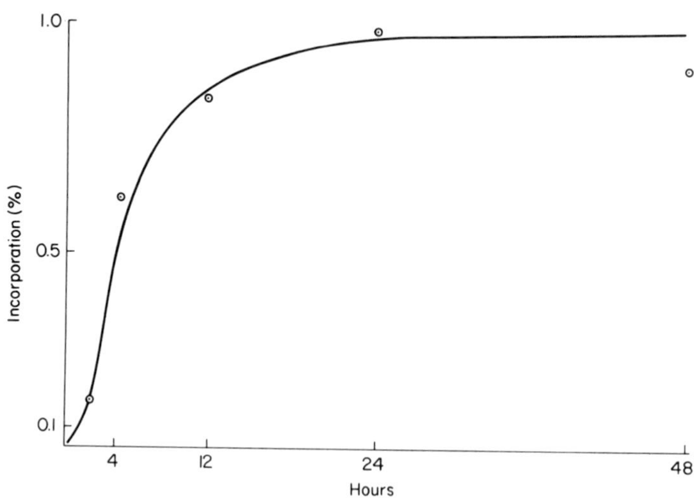

FIG. 5. Incorporation of DL-tryptophan-^{14}C into alkaloids in 3-month-old shoots of *Vinca rosea* during 48 hours.

concluded from these results that there is present no large pool of tryptophan and also that, since after 8 days vindoline (**18**) (a major alkaloid of *V. rosea*) contains only 7 percent of all the activity, 8 days is insufficient time for the distribution of radioactivity to approximate to the "normal" percentage of alkaloid distribution in the seedlings. Shoots of mature *V. rosea* which have been hitherto used for most of the published feeding data incorporate tryptophan but at a reduced initial rate, and the final (constant) percentage is about 1.0 (Fig. 5). The latter observation is of particular interest for comparative studies with diversely aged biological material. When the same techniques were applied to the uptake of MVA-2-^{14}C almost 90 percent of the label was incorporated into ursolic acid, the major triterpene of *V. rosea*. It is assumed that the iridoid pool size is so large in the seedlings that the method offers little advantage for studying this aspect.

From Vincoside to the Corynanthé Alkaloids

In spite of considerable progress in the determination of the sequence of events in *V. rosea* whereby tryptophan and secologanin are condensed and transformed to both simple and complex members of the three major families of alkaloid (Scott, 1970), autoradiograms of two-dimensional TLC systems disclose (Fig. 6) that the earliest alkaloid into which tryptophan is incorporated is (as yet) an unknown base. Thus at the earliest sampling time (5 minutes) this metabolite contains 35 percent of the total counts of the alkaloidal fraction. Within 1 hour the unidentified compound contains virtually no radioactivity. These results are consistent with the view that the plot is characteristic of an active intermediate with a constant pool size corresponding to 35 percent of the radioactivity at 5 minutes, at which time a linear incorporation into the alkaloids is being maintained (Figs. 3 and 4). Comparison of the data of Fig. 6 with the profile for geissoschizine (**5**) shows (Fig. 7) that the unknown lies between vincoside (**1**) (not extracted from aqueous solution) and geissoschizine (**5**). The nearest (but nonidentical) alkaloid (R_f in two solvent systems) is corynantheine aldehyde (**6**).

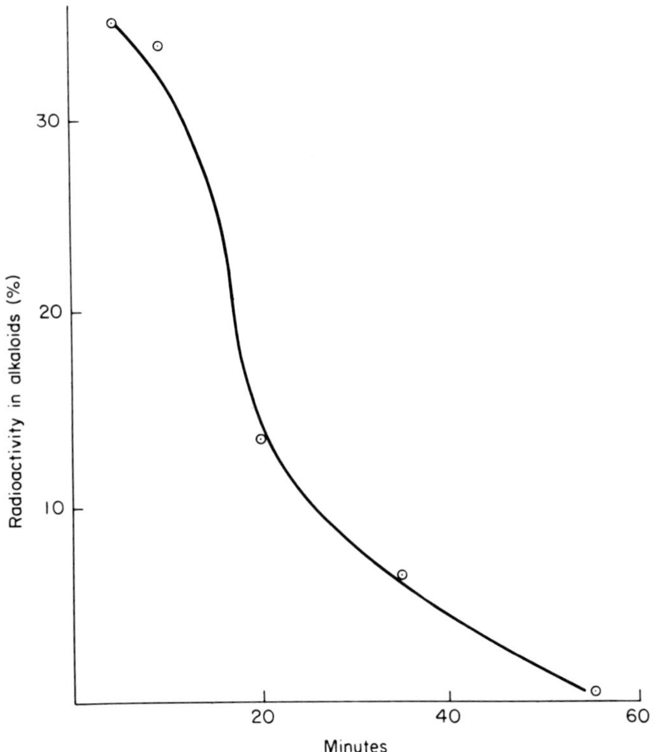

FIG. 6. Radioactivity of unknown alkaloid (t = 5–6 minutes) of *Vinca rosea* after administration of DL-tryptophan-^{14}C to seedlings.

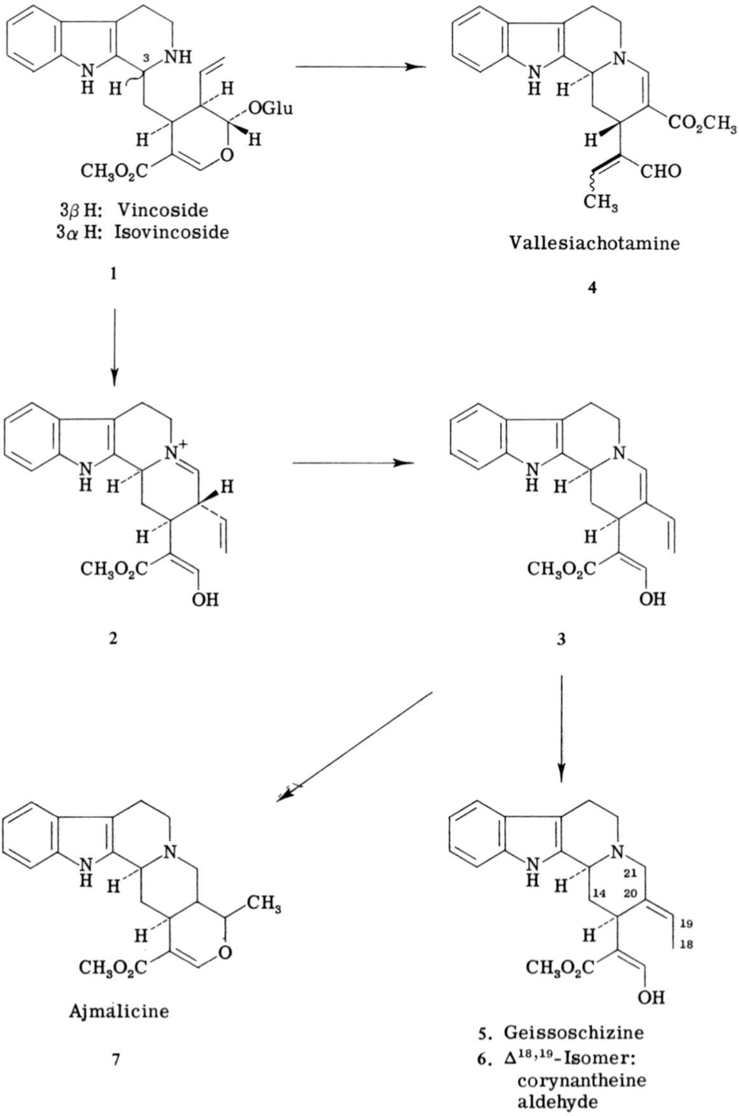

3β H: Vincoside
3α H: Isovincoside

1

Vallesiachotamine

4

2

3

Ajmalicine

7

5. Geissoschizine
6. $\Delta^{18,19}$-Isomer: corynantheine aldehyde

SCHEME 1. Vincoside and corynanthé alkaloids.

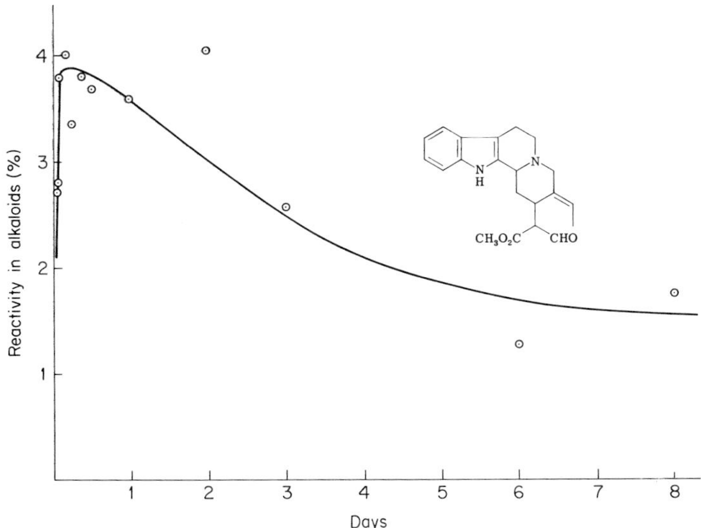

FIG. 7. Radioactivity of geissoschizine during 8 days after administration of DL-tryptophan-^{14}C.

A second possibility that the new intermediate might be vallesiachotamine (**4**) could also be ruled out by TLC comparison with an authentic sample (Djerassi *et al.*, 1966). Since it had been shown that vincoside (strictosidine) is transformed to vallesiachotamine during the normal work-up procedure (Smith, 1968), it seemed possible that the appearance and disappearance of vincoside (**1**) from the system could be monitored in this way. In fact, the recent revision by Pinar *et al.* (1971) of the stereochemistry at C-3 in **1**, for that epimer which is incorporated into the main classes of alkaloids, appears to be in accord with our observation, although the time scale may have to be shortened still further (1–250 seconds) before such species as vallesiachotamine can be eliminated completely from consideration.

Another attractive candidate for the unidentified pre-corynanthé alkaloid is the unknown diene, $\Delta^{20:21}$-corynantheine aldehyde (**3**). The structure of this dieneamine represents a link between the highly reactive aglucone of vincoside, the corresponding immonium species (**2**) and (by enamine–imine tautomerism) the dehydro version (**5**; $\Delta^{N:21}$) of geissoschizine. This would imply an indirect connection between the corynantheine ($\Delta^{18,19}$) and geissochizine ($\Delta^{19,20}$) series and is in accord with the similar but non-identical TLC behavior of the new intermediate and that of corynantheine aldehyde (**6**). So far, only microgram quantities of this alkaloid are available and further isolation experiments are in hand.

Turning now to Figs. 7 and 8, it can be seen that within the limits of the technique the next detectable intermediate is geissoschizine (**5**) which bears a profile expected from previous feeding experiments (Scott, 1970), i.e., a rapid rise to approximately 4 percent of total radioactivity in the alkaloids over the first 60–90 minutes followed by a gradual decline over the 8-day experiment (Fig. 7). This curve is to be contrasted with the behavior of ajmalicine (**7**) (Fig. 9), which although isomeric with geissoschizine no longer exhibits the required dynamic structure for further transformation, a fact reflected in its rather slow climb and flat profile over the 8-day incubation.

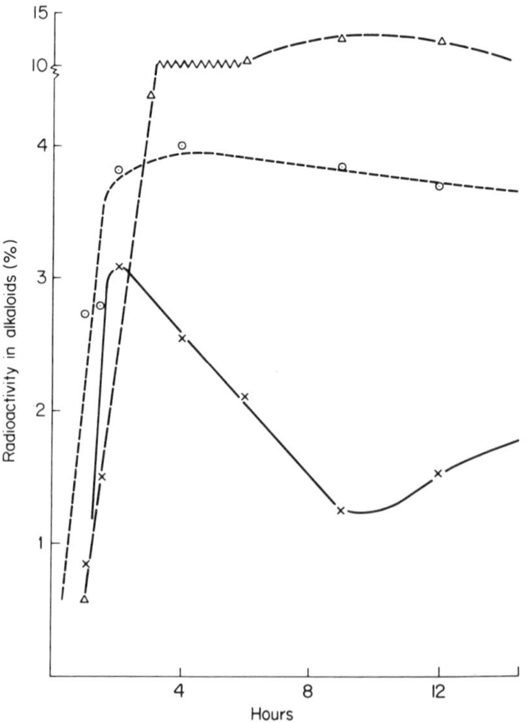

Fig. 8. Relative radioactivities in geissoschizine (— ⊙ —), akuammicine (— × —), and tabersonine (— Δ —) over 12 hours.

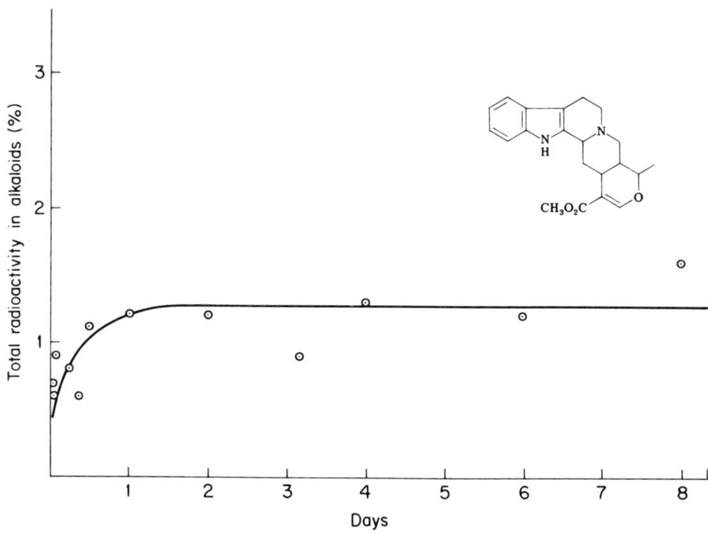

FIG. 9. Radioactivity in ajmalicine after administration of DL-tryptophan-^{14}C during 8 days.

The Corynanthé-Strychnos Relationship

Previous work in these laboratories (Scott, 1970) demonstrated that geissoschizine serves as a good precursor both for akuammicine (**10**) (*Strychnos*) and coronaridine (**17**) (*Iboga*). When the profile of the akuammicine was examined (Figs. 8 and 10), it at first seemed surprising that a definite maximum (3 percent) occurs after about 2 hours. Yet thereafter the amount of radioactivity in both akuammicine and its phenolic derivative vinervine (**11**) remains virtually unchanged from 9 hours to 8 days. The latter part of the akuammicine profile suggests that both **10** and **11** are shunt products formed from a rapidly metabolizing intermediate at a constant rate during the early part of the biosynthesis. The rapidly changing segment of the curve (0–9 hours) then may well represent the flow of radioactivity through the labile intermediate preakuammicine (**8**) (Scott, 1970). It is known from the chemistry of this compound that, under the normal conditions of work-up, almost quantitative conversion to akuammicine occurs. Thus the first 9 hours of the experiment may constitute an assay for preakuammicine which is analyzed as the nor-derivative, akuammicine. The latter of course does not serve as a mainstream intermediate but is presumably the precursor of vinervine (**11**).

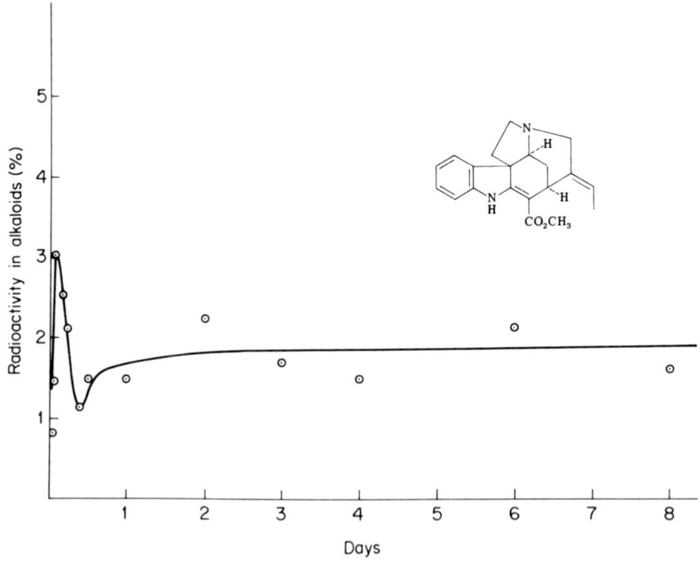

FIG. 10. Radioactivity in akuammicine from feeding DL-tryptophan-^{14}C for 8 days.

The next isolated alkaloid of *V. rosea* which (like geissoschizine) has been demonstrated to give rise to both the *Aspidosperma* and *Iboga* alkaloids is stemmadenine (**9**). Stemmadenine and preakuammicine are related by oxidation of the former at position 3 while reduction of preakuammicine has been used to illustrate the reversal of this process in the laboratory. In spite of the successful, nonrandomized bioconversion of stemmadenine to both vindoline (**18**) and catharanthine (**16**) in *V. rosea* (Scott, 1970) and to related alkaloids in *V. minor* (Kutney *et al.*, 1969, 1971a), the radioprofile for stemmadenine (Fig. 11) does not reveal any dramatic rise in radioactivity at the expected time interval, i.e., between preakuammicine (2 hours) and tabersonine (9 hours). The possibility exists that the preakuammicine–stemmadenine equilibrium is an enzyme-bound process which precludes analysis of a dynamic intermediate such as stemmadenine.*

* An alternative interpretation of the stemmadenine profile implicates this intriguing alkaloid as a reduced, stabilized version of preakuammicine (**8**), which (especially in young seedlings, Scott, 1970) can be brought into the biogenetic scheme by oxidative cyclization (**9** → **8**). Bioconversion and isolation studies (Scott, 1970) leave no doubt that the "normal" criteria of intermediacy are satisfied. However, the radioprofile [together with the extremely low (0.001%) incorporation of tryptophan-^{14}C into stemmadenine] serves as a third criterion which is not met in the dynamic sense by compound **9** under our experimental conditions. The answer to this problem must await development of the cell-free technique with *V. rosea*.

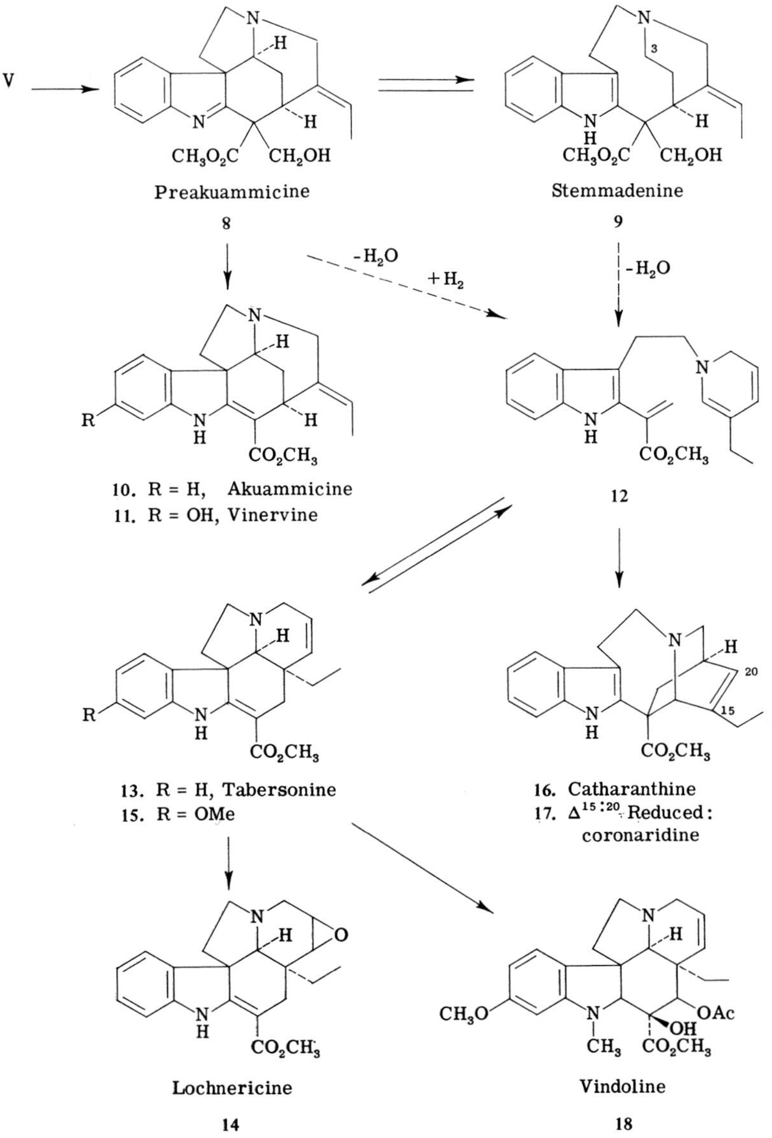

V →

CH_3O_2C CH_2OH

Preakuammicine

8

CH_3O_2C CH_2OH

Stemmadenine

9

$-H_2O$ $+H_2$

$-H_2O$

CO_2CH_3

10. R = H, Akuammicine
11. R = OH, Vinervine

CO_2CH_3

12

CO_2CH_3

13. R = H, Tabersonine
15. R = OMe

CO_2CH_3

16. Catharanthine
17. $\Delta^{15:20}$ Reduced: coronaridine

CO_2CH_3

Lochnericine

14

CH_3O OAc OH CH_3 CO_2CH_3

Vindoline

18

SCHEME 2. Strychnos, aspidosperma, and iboga alkaloids.

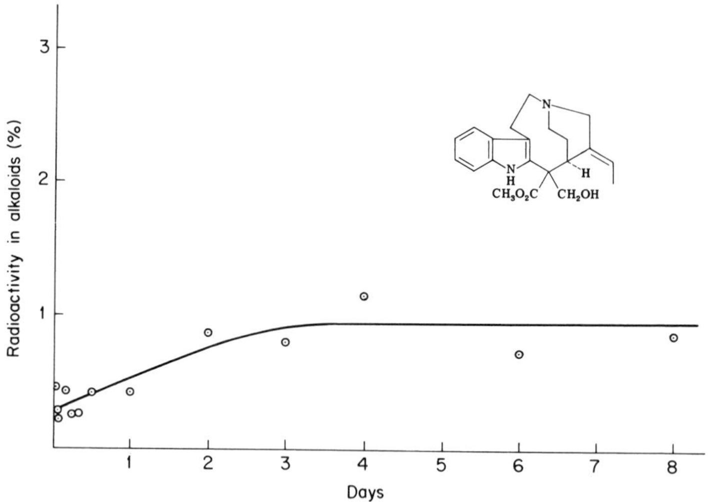

FIG. 11. Radioactivity in stemmadenine during 8 days.

In the same category may be included the next plausible intermediate, *viz.* the labile acrylic ester (**12**) which has served as a satisfactory mechanistic concept to connect the preakuammicine series with *both Aspidosperma* and *Iboga* alkaloids and for which a considerable body of circumstantial evidence now exists in terms of its generation both *in vivo* (Scott, 1970) and *in vitro* (Scott and Cherry, 1969), as we shall discuss in a later section.

Aspidosperma and *Iboga* Alkaloids

One of the earliest isolation experiments in which new metabolites of immature *V. rosea* were identified concerned the characterization of (—)-tabersonine (**13**) as an abundant constituent of 3-day-old seedlings, an observation which contrasts with the virtual absence of tabersonine in the mature plant. The possibility was tested that tabersonine (**13**) (the simplest of the *Aspidosperma* alkaloids) could serve not only as a precursor for the more abundant *V. rosea* constituent, vindoline (**18**) and related highly oxygenated *Aspidosperma* alkaloids, but also as a precursor of the *isomeric Iboga* alkaloid catharanthine (**16**) once again via the *chano*-intermediate (**12**). Since tabersonine and catharanthine are formed at different rates (they are isolated at 72 and 120 hours, respectively) from geissoschizine, this implied a special role for tabersonine in *V. rosea* and hence in many other species as a prototype of *Aspidosperma*, *Iboga*, and *Hunteria* bases. The profile of tabersonine reveals that (in accord with these expectations) a remarkably dynamic role must indeed be ascribed to the alkaloid. Thus after 9 hours the tabersonine pool has attained maximum radioactivity (Fig. 8). At this time tryptophan-2-^{14}C of specific activity 52 mCi/mmole gave rise to tabersonine of specific activity 15.6 mCi/mmole (30 percent specific incorporation), indicating the extremely small pool size of the tryptophan available for alkaloid synthesis and of tabersonine. Autoradiograms clearly showed that tabersonine was metabolized almost as rapidly as it was formed, and in fact its activity falls (Figs. 12 and 13) from 12 percent to 2 percent in 3 days. This behavior is similar to that of the unknown alkaloid (Fig. 6), but is complicated by the falling rate of incorporation of tryptophan into the alkaloids after 1 day. Refeeding ^{14}C-labeled tabersonine isolated after feeding tryptophan for 9 hours gives a number of labeled metabolites. After 1 day when there is still a large amount of unchanged tabersonine (**13**), it is found that epoxytabersonine (**14**), methoxytabersonine (**15**), and coronaridine (**17**) are all labeled. After 6 days there is almost no tabersonine remaining while vindoline (**18**) has gained activity (Fig. 12).

The role of epoxytabersonine [(−)-lochnericine] (**14**) in the overall biosynthetic map is at present obscure. It is a minor constituent of the normal alkaloid pattern of *V. rosea*, but after 9 hours it has 4.3 percent of all the

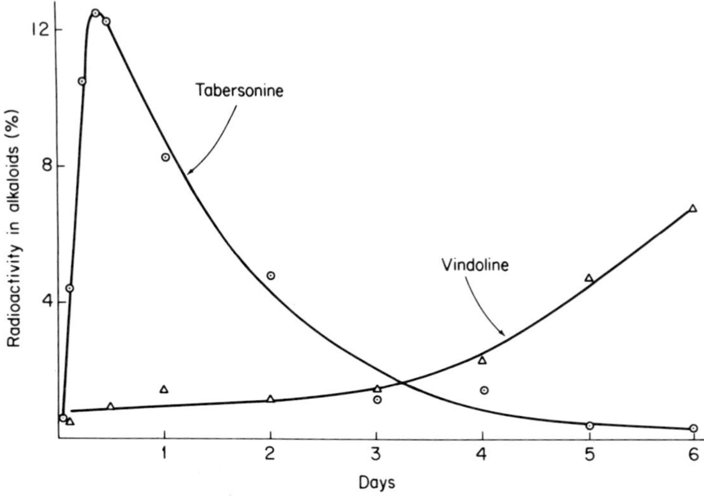

Fig. 12. Relative radioactivities in tabersonine (— ⊙ —) and vindoline (— Δ) during 6 days.

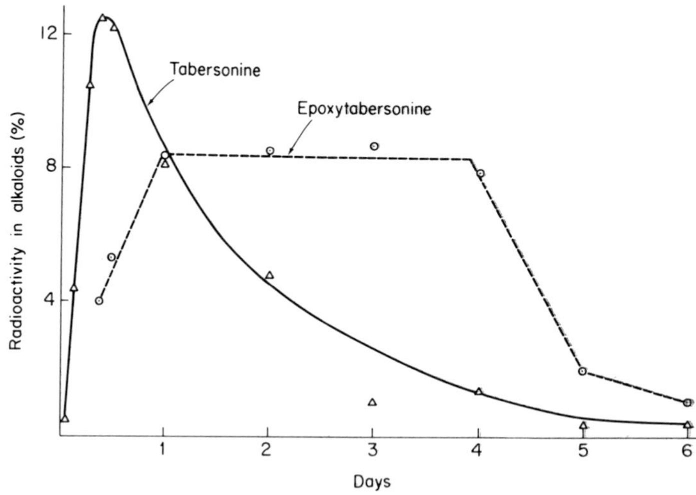

FIG. 13. Radioactivities in tabersonine (— Δ) and its epoxide lochnericine (— ⊙ —) over 6 days.

activity (Fig. 13), a specific activity of 10 mCi/mmole (i.e., 20 percent specific incorporation) and is formed from tabersonine as shown from refeeding the latter labeled with ^{14}C from tryptophan-^{14}C. Refeeding (**14**) gave rise to several unidentified alkaloids.

Although vindoline can be detected on autoradiograms after 1 day, its activity does not begin to increase rapidly until after 3 days (Fig. 12). At this stage the amount of radioactivity in the alkaloids has reached a maximum, and the activity in vindoline will presumably increase rapidly to a maximum value that will not alter unless vindoline is broken down.

Soon after the onset of the decline in radioactivity in tabersonine, the principle *Iboga* alkaloid catharanthine (**16**) is labeled, and a linear increase is observed from day 1 to day 8 to reach a value of 6 percent of the radioactivity in the alkaloids at the end of the experiment (Fig. 14). The dihydro form of catharanthine, coronaridine (**17**), exhibits a fluctuating profile (Fig. 15) which may reflect a control by its (presumed) precursor, catharanthine. Complementary experiments with tritiated versions of catharanthine and coronaridine indicate (in the whole plant) that there is no evidence for a connection between **16** and **17** in *either direction*. This surprising result implies a separate biosynthetic pathway for two closely related *Iboga* alkaloids but may be rationalized by the stereospecific labeling experiments described below.

In summary, a set of dynamic precursors has been detected by short-term incubation. Several of these, but not all, correspond to previously discovered intermediates. In addition to the unknown alkaloid which is maximally labeled after 5 minutes, the appearance of the "radioprofiles" of geissoschizine (**5**) (1.5–2 hours), preakuammicine (**8**) (2.5–2 hours), and tabersonine (**13**) (9 hours) distinguish these alkaloids from ajmalicine (**7**), catharanthine (**16**), coronaridine (**17**), lochnericine (**14**), and vindoline (**18**). Confirmation of the intermediacy of the former set has been secured for compounds **5** and **13** in separate feeding experiments (Scott, 1970). The specific incorporations of tryptophan into the alkaloids reaches a most satisfactory level in these experiments, e.g., tabersonine (30 percent) and its epoxide (20 percent). Again the precursor role of tabersonine not only for the *Aspidosperma* but also for the *Iboga* series is confirmed. The mechanism of the latter process may in fact involve the generation of both (+) and (−) forms of tabersonine, one of which [the (−) form] leads to vindoline, the other, hitherto unknown (+)-form undergoing transformation to catharanthine and/or coronaridine. The details of this mechanism and the stemmadenine–preakuammicine–acrylic ester relationship are discussed below.

Application of the technique to several other putative precursors and the

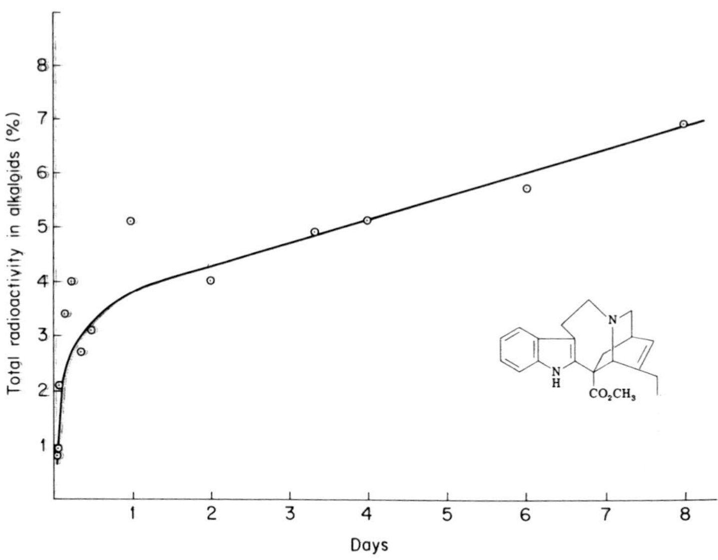

FIG. 14. Radioactivity in catharanthine (8 days).

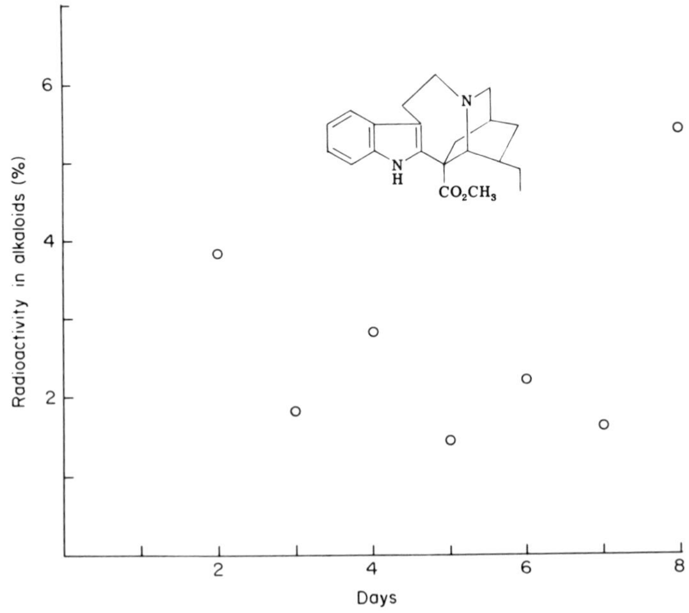

Fig. 15. Radioactivity in coronaridine (8 days).

development of rapid sampling methods to evaluate a shorter time scale (0–300 seconds) with substrates of very high specific activity are in progress.

Rearrangement of the Mevalonate-Derived Segment

Throughout the foregoing discussion we have assumed that the precursor relationship of the indole alkaloids of *V. rosea* (and by implication of other species) follows the sequence: *Corynanthé* → *Aspidosperma* → *Iboga*. All of the published incorporation data are in accord with this idea, but so far we have not considered in intimate detail how these profound skeletal changes might come about except to comment on the more obvious (but unproved) steps connecting vincoside (**1**) with geissoschizine (**5**). The conversion of geissoschizine (**5**) to the prototype of all the other families, preakuammicine (**8**), may involve the recently isolated geissoschizine oxindole via mechanisms already discussed in some detail elsewhere (Scott, 1970) and summarized in Fig. 16. Deformylation of preakuammicine to the *Strychnos* alkaloid, akuammicine (**10**), is unexceptional and has been demonstrated *in vitro*, as has the reductive conversion of **8** to stemmadenine (**9**). The reversal of this latter process using Pt–O_2 has also been achieved (Scott, 1970). These experiments were designed as part of a search for ways of generating the elusive acrylic ester (**12**), which in our view provides the key to the most intriguing problem of this complex system, namely the question of how a common intermediate such as preakuammicine or stemmadenine which, in view of the profile discussed above, may well represent an equilibrating shunt process, can rearrange to *both* the *Aspidosperma* and *Iboga* templates. In this connection we may mention several indirect pieces of evidence in accord with our original postulate that the acrylic ester (**12**) is an attractive but fugitive biointermediate. The recent

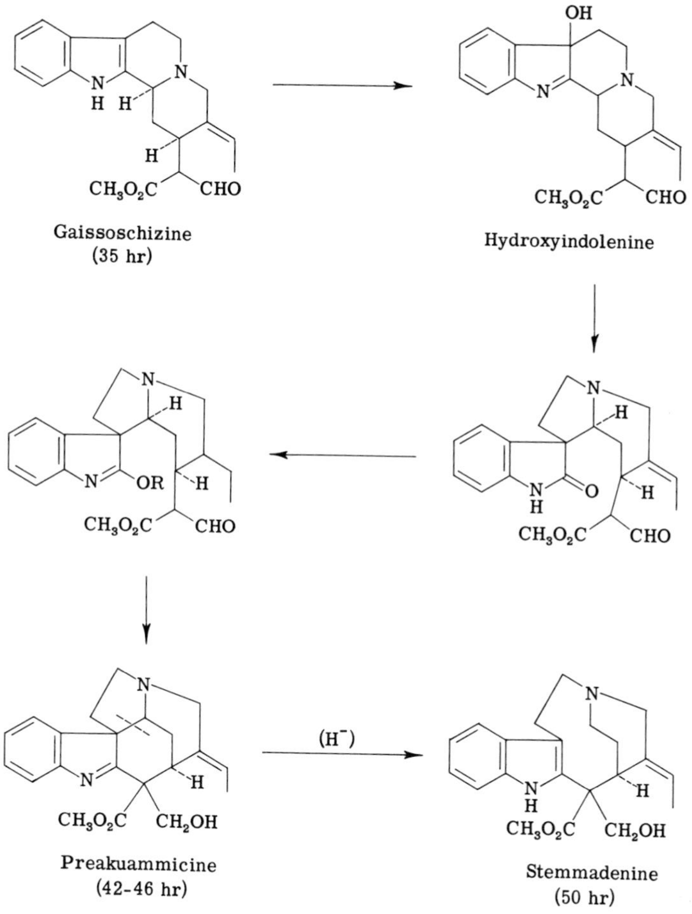

Gaissoschizine (35 hr)

Hydroxyindolenine

Preakuammicine (42-46 hr)

8

(H^-)

Stemmadenine (50 hr)

9

FIG. 16. Proposed mechanism for the geissoschizine → stemmadenine pathway.

isolation of members of the secodine family (Battersby and Bhatnagar, 1970) indicates that such a *chano* intermediate could be generated in nature from stemmadenine (⇆ preakuammicine), or perhaps even tabersonine or catharanthine by cleavage of the appropriate carbon–carbon bonds. Not only can these bonds be severed by the enzymes of *V. rosea,* but a striking series of *in vitro* models provide excellent analogy for the production of not only the *secodines* but also *Aspidosperma* alkaloids. Thus based on preliminary experiments (Fig. 17) in which the *racemic* versions of *pseudo*-catharanthine and tabersonine, the main products of a low-yielding, capricious, but nevertheless promising reaction of stemmadenine, further conditions were sought (Scott and Cherry, 1969), which could be used not only as a model for biosynthesis but as a useful one-step synthetic operation.

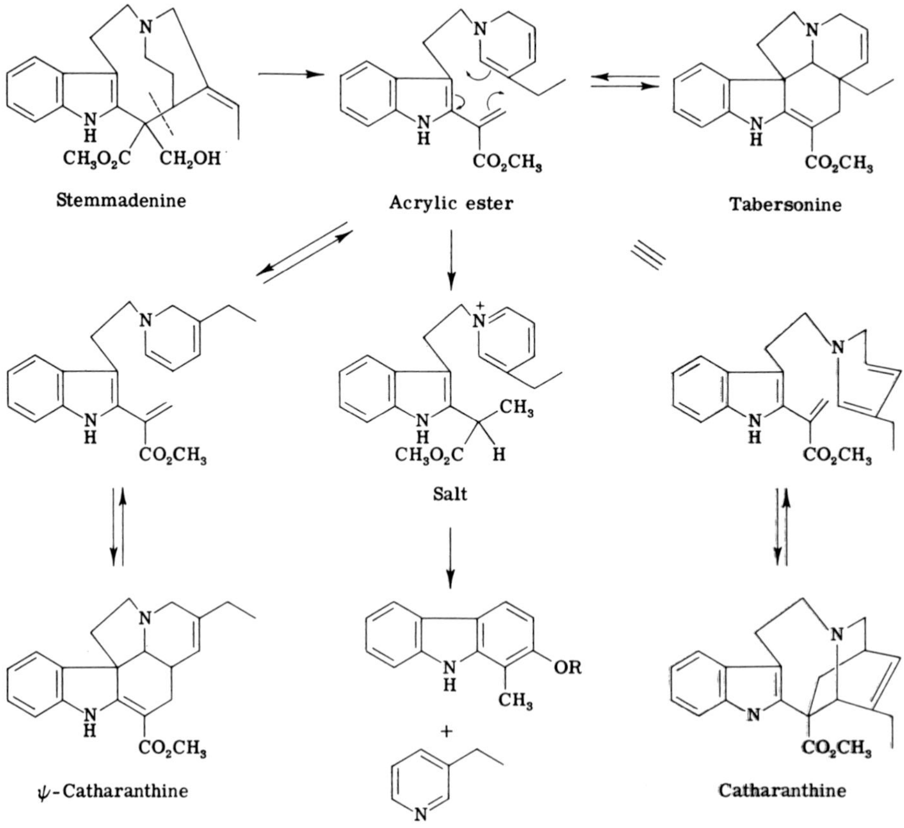

FIG. 17. *In vitro* models for acrylic ester biogenesis.

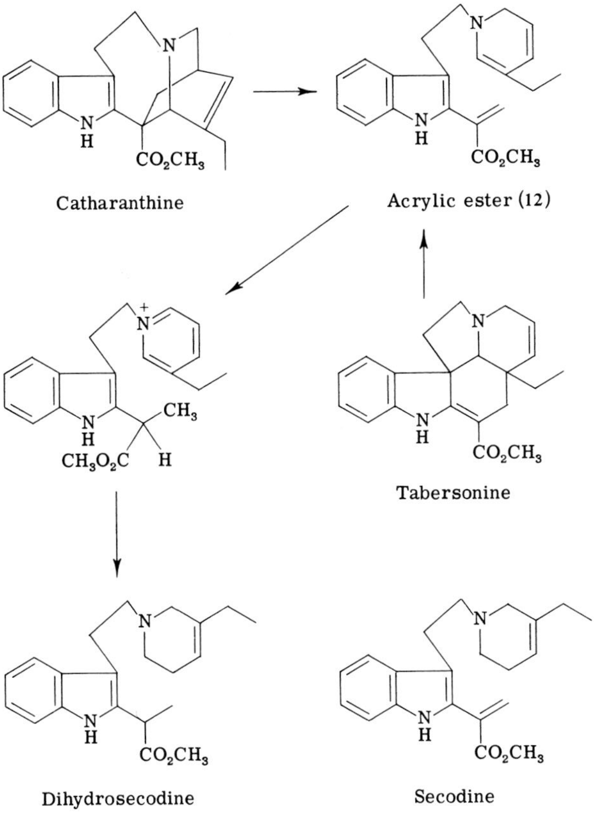

Catharanthine

Acrylic ester (12)

Tabersonine

Dihydrosecodine

Secodine

19

FIG. 18. Biogenetic type synthesis of secodines.

After considerable experimentation it was possible to demonstrate, for example, the conversion of 50 percent yield of catharanthine in methanol solution and the seco-salt (Fig. 18), which in turn serves as a useful source of the crystalline racemic dihydrosecodine (**19**) via borohydride reduction. The latter compound is identical in every respect with the amorphous natural material isolated from *Rhazya stricta* in milligram quantities (Brown *et al.*, 1970; Cordell *et al.*, 1970). Furthermore the acrylic ester model has been used in an excellent stereospecific synthesis of *Aspidosperma* alkaloids by Ziegler and Spitzner (1970). Recently the bioconversion of secodine to vindoline has been reported by Kutney *et al.* (1971b). It is therefore somewhat surprising to find that a criticism (Smith *et al.*, 1969) of our original report (Qureshi and Scott, 1968) of the rearrangement chemistry of stemmadenine and tabersonine, which in our hands depended on the generation of *racemic* materials (as determined by ORD measurement), rests largely upon assays which in the authors' words were performed with preparations of "pseudocatharanthine whose optical activity, (−) or (±)" is unknown. Since *pseudo*-catharanthine is a well defined compound existing in crystalline form as a *racemate,* we can only conclude that all the tests for the alleged failure of the reactions described earlier by us were carried out with material *which is not identical with the reference compound.* Thus the "prolonged and careful study" which has cast doubt on the original report should be regarded with reserve until the reaction conditions have been reevaluated scientifically. Very recently a new source of stemmadenine has allowed these aspects to be reinvestigated. However, we note at this time, as a result of our preliminary finding, that all the mechanisms implicit in our theory have been demonstrated (Scott and Wei, 1972) good synthetic methods have been evolved from the principles established, and a number of compounds related to the acrylic ester (**12**) have been isolated from plant sources.

Returning to the biosynthetic mechanism, our experience in generating the acrylic ester (Fig. 18) indicated that an indirect method would be required to secure proof for its intermediacy. To this end, consideration was given to the fate of the protons of MVA as they pass from the *Corynanthé* to the *Aspidosperma* family via the acrylic ester. In particular, if we trace the pro 5(*R*) and 5(*S*)-MVA protons through the established

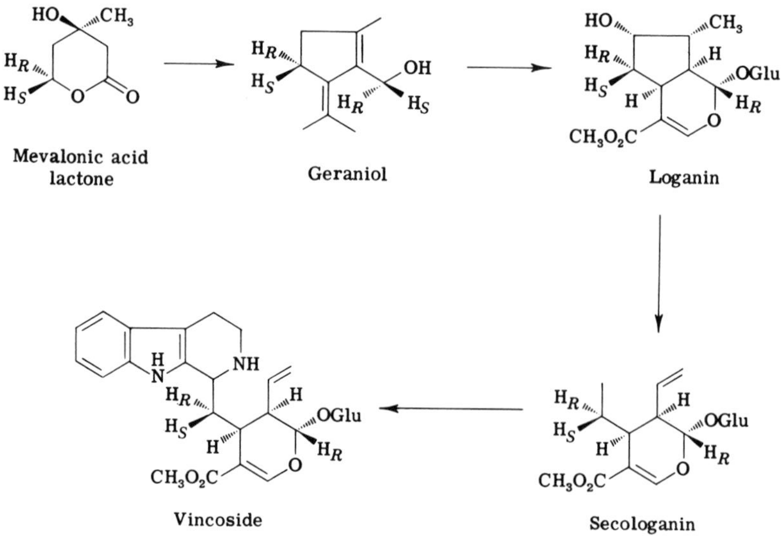

Fig. 19. Conversion of MVA to secologanin and vincoside.

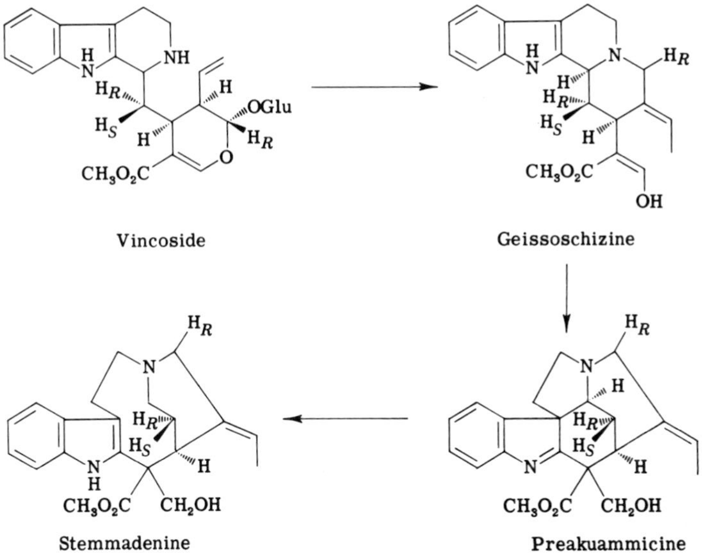

Vincoside → Geissoschizine → Preakuammicine → Stemmadenine

FIG. 20. Conversion of vincoside to stemmadenine.

precursors: geraniol ($2H_R{:}2H_S$), loganin ($2H_R{:}1H_S$), and vincoside ($2H_R{:}1H_S$) (Fig. 19), we should not expect any further loss or change to occur until after the preakuammicine–stemmadenine stage (Fig. 20). However when the acrylic ester is generated (Fig. 21) various possibilities are open. As can be seen from Table 1, administration of MVA-5(R,S)-^{3}H-5-^{14}C to *V. rosea* leads to a sample of loganin which has retained 3 of the

TABLE 1

ADMINISTRATION OF STEREOSPECIFICALLY LABELED MEVALONIC ACIDS TO *Vinca rosea*[a]

^{3}H atoms retained (2 units of MVA)	MVA-5(RS)-^{3}H-5-^{14}C	MVA-5(S)-^{3}H-5-^{14}C
Loganin[b]	2.9 (3.0)	—
Vindoline	1.5, 1.8 (2.0)	0.25 (0.0)
Catharanthine[c]	2.1, 2.5[d] (2.0)	0.81 (1.0)

[a] "Theoretical" values in parentheses.
[b] Pentaacetate.
[c] HCl^{ide}
[d] ^{3}H isotope effect.

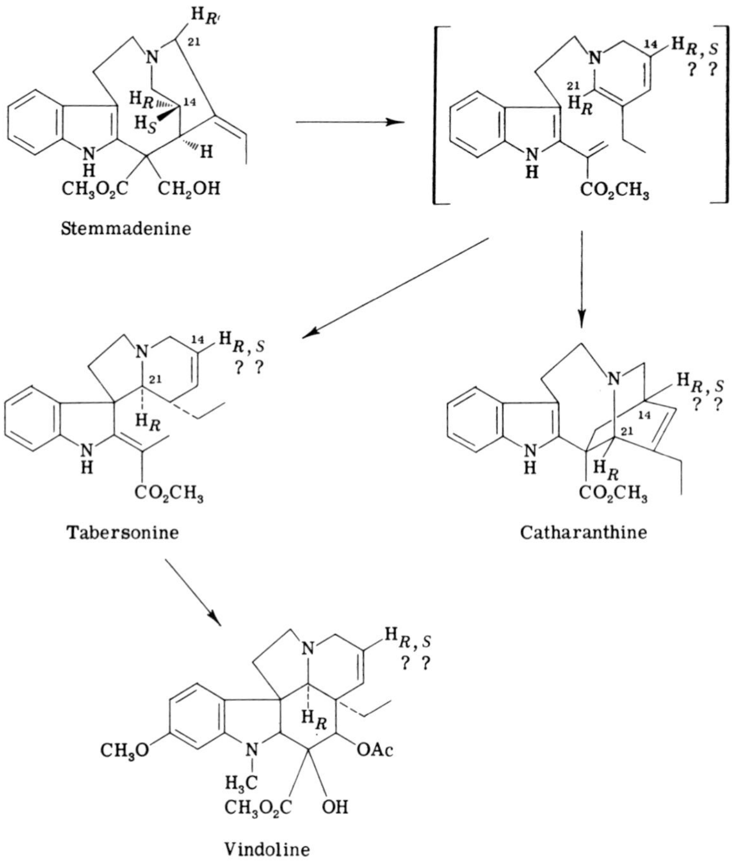

Stemmadenine

Tabersonine

Catharanthine

Vindoline

FIG. 21. Conversion of stemmadenine to vindoline and catharanthine.

original 4 protons of the two MVA units. In accord with the mechanism in Fig. 20, both vindoline and catharanthine retain *two* of these protons (Table 1). As indicated in Fig. 20, this experiment does not disclose which protons (R or S) are lost in the rearrangement steps. In order to make this distinction, a new version of MVA was synthesized using rat liver mevaldate reductase and synthetic tritiated aldehyde. The details of the synthesis and proof of absolute stereochemistry have been described elsewhere (Scott *et al.*, 1970). When this species of MVA-5(S)-^{3}H, admixed with the 5-^{14}C-radiomer as internal standard, was fed to shoots of *V. rosea* ,the isolated vindoline and catharanthine no longer contained identical ^{3}H:^{14}C ratios (Table 1). In fact the results show beyond any doubt that while vindoline (and hence tabersonine, its direct intermediate) has lost almost all the original 5(*S*)-^{3}H, catharanthine still contains one 5(*S*)-proton (within experimental error).

In terms of a simple theory for the generation of the acrylic ester from tabersonine, an established precursor of coroanaridine and catharanthine, this result provides a challenge of interpretation, for there seems at first sight no way for a specimen of (−)-tabersonine (H_R, H_R) (Fig. 21) to be bioconverted to catharanthine (H_R, H_S). A possible solution to this apparent paradox may be developed as follows.

The acrylic ester (Fig. 22) is generated from stemmadenine with retention of the three C_5-MVA protons ($2H_R$:H_S). The dihydropyridinium system then loses either the H_S proton to give the dienamine ($2H_R$) and thence (−)-tabersonine and vindoline with the observed label (H_R, H_R). If the H_R proton is lost in an enantiotopic process, the resultant acrylic ester can form either catharanthine (H_R, H_S) or by virtue of its enantiomeric conformation generate a new alkaloid (+)-tabersonine (H_R, H_S), which in turn suffers ring opening to that form of the acrylic ester which can recyclize to to catharanthine (H_R, H_S).

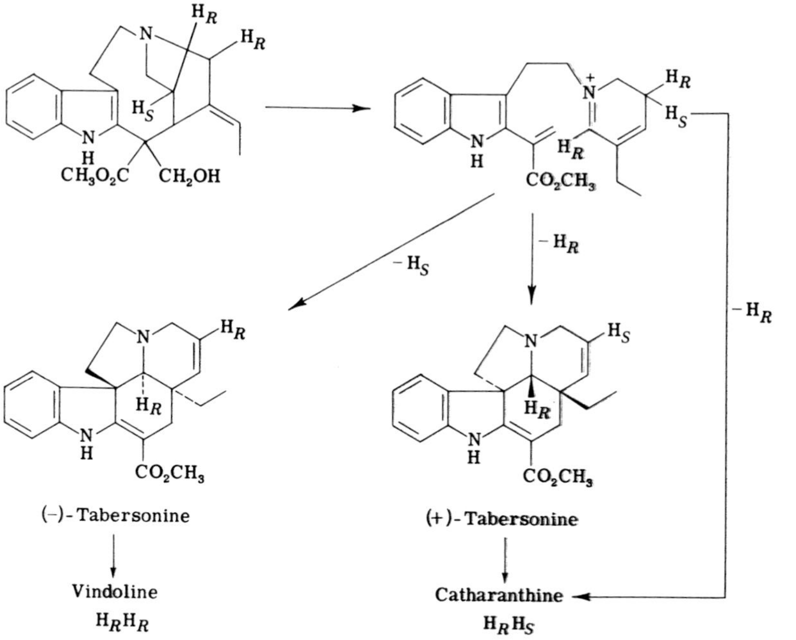

Fɪɢ. 22. Fate of prochiral protons in stemmadenine and the acrylic ester.

TABLE 2

RECORDED [α]D VALUES FOR SOME *Aspidosperma* β-AMINO ACRYLATE ALKALOIDS[a]

(−)-Tabersonine	−310°, −380°, −366°, −294°
Dihydrotabersonine (vincadifformine)	−540°, 0°, +600°
Epoxytabersonine	−432°, −528°, −36° (!)
(+)-Echitovenin	+640°
(−)-Minovincin	−534°
(±)-Minovine	0°

[a] From Hesse (1964, 1968).

The intervention of (+)-tabersonine is suggested in order to explain the independent reports from two laboratories (see Scott, 1970) that "natural" tabersonine is biotransformed to catharanthine. Since optically pure (−)-tabersonine (H_R,H_R) would have lost all the H_S label which appears in catharanthine, our explanation of what appears to be a conflicting result is that "natural" (−)-tabersonine used in the feeding experiment contains a small percentage of the (+)-enantiomer which is responsible for the formation of catharanthine. Indirect support for this idea comes from (a) the recorded occurrence of dihydrotabersonine (vincadifformine) in (+), (−), and (±) forms as well as the rather low and variable $[\alpha]_D$

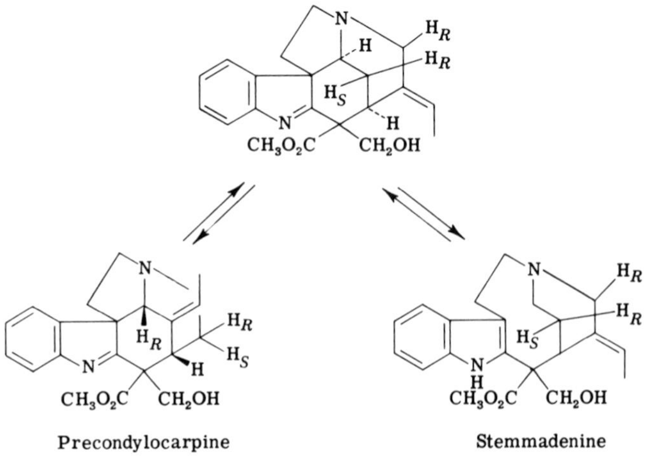

Precondylocarpine

Stemmadenine

FIG. 23. Stemmadenine in equilibrium with preakuammicine and precondylocarpine.

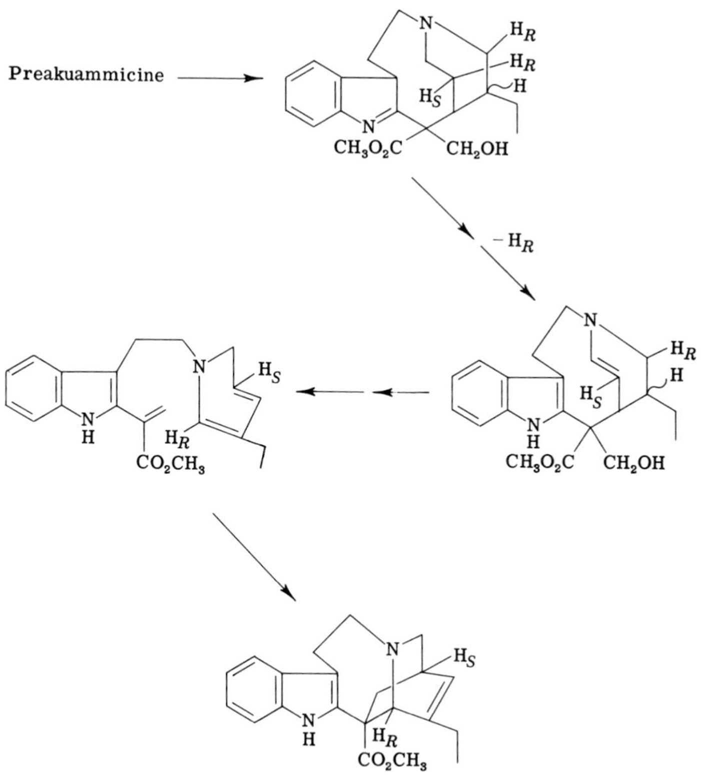

Preakuammicine

Catharanthine

FIG. 24. The preakuammicine-catharanthine series (hypothetical).

reported for (−)-tabersonine compared with other members of the series (Table 2), which all contain the β-aminoacrylate chromophore and differ only in oxidation level at a point remote from the main chromophore responsible for the optical rotatory power, and (b) the ratio of incorporation (>10:1) of tabersonine into vindoline compared with catharanthine.

An alternative explanation of the difference in labeling between the *Aspidosperma* and *Iboga* alkaloids involves variants of quite complex schemes (Figs. 24 and 25) which reflect the biogenetic-type chemistry of stemmadenine (Scott and Wei, 1972). Thus a distinction between the H_R and H_S protons could be made if stemmadenine is in equilibrium with preakuammicine and precondylocarpine (Fig. 23). Reduction of these species could lead in time to the labeling patterns depicted in Figs. 24 and 25, respectively. A less complicated version of this process is the stereospecific

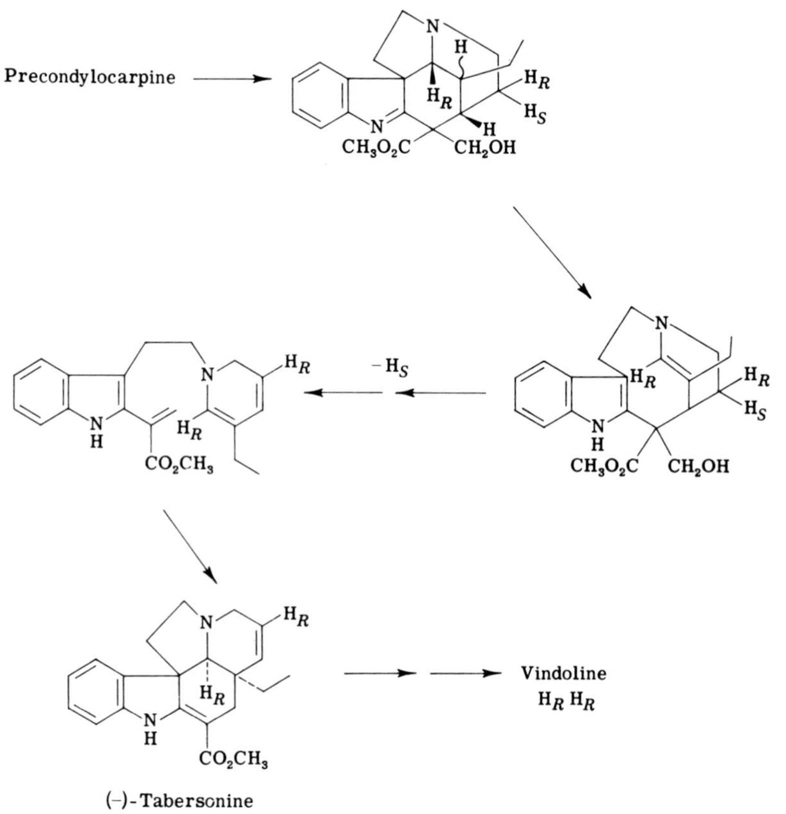

Precondylocarpine → [structure] → [structure] —H_S→ [structure] → (−)-Tabersonine → → Vindoline H_R H_R

FIG. 25. The precondylocarpine-vindoline series (hypothetical).

dehydrogenation of stemmadenine in which either the H_R or H_S proton in question is lost.

Further work to test all these hypotheses is in progress.

ACKNOWLEDGMENT

This work was supported by grants from the National Institutes of Health and the National Science Foundation.

REFERENCES

Banthorpe, D. V., and A. Wriz-Justice. 1969. *J. Chem. Soc. C* p. 541.
Battaile, J., A. J. Burbott, and W. D. Loomis. 1968. *Phytochemistry* 7:1159.
Battersby, A. R., and A. K. Bhatnagar. 1970. *Chem. Commun.* p. 193.
Brown, R. T., G. F. Smith, K. S. J. Stapleford, and D. A. Taylor. 1970. *Chem. Commun.* p. 190.

Cordell, G. A., G. F. Smith, and G. N. Smith. 1970. *Chem. Commun.* p. 189, 191.
Djerassi, C., H. J. Monteiro, A. Walser, and L. J. Durham. 1966. *J. Amer. Chem. Soc.* **88**:1792.
Fales, H. M., J. D. Mann, and S. H. Mudd. 1963. *J. Amer. Chem. Soc.* **85**:3820.
Hesse, M. 1964. "Indole Alkaloids in Tables." Springer-Verlag, Berlin and New York.
Hesse, M. 1968. "Indole Alkaloids in Tables." Springer-Verlag, Berlin and New York.
Kutney, J. P., V. R. Nelson, and D. C. Wigfield. 1969. *J. Amer. Chem. Soc.* **92**:4278.
Kutney, J. P., J. F. Beck, V. R. Nelson, and R. S. Sood. 1971a. *J. Amer. Chem. Soc.* **93**:255.
Kutney, J. P., J. F. Beck, C. Ehvet, G. Poulton, R. S. Sood, and N. D. Wescott. 1971b. *Bioorg. Chem.* **1**:194.
Mudd, S. H. 1960. *Biochim. Biophys. Acta* **38**:354.
Pinar, M., M. Hanaoka, M. Hesse, and H. Schmid. 1971. *Helv. Chim. Acta* **54**:15 (quoting A. R. Battersby).
Qureshi, A. A., and A. I. Scott. 1968. *Chem. Commun.* p. 945.
Rapoport, H., F. R. Stermitz, and D. R. Baker. 1960. *J. Amer. Chem. Soc.* **82**:2765.
Robinson, Sir R. 1917. *J. Chem. Soc., London* **111**:876.
Robinson, Sir R. 1955. "The Structural Relations of Natural Products." Oxford University Press, London and New York.
Schechter, I., and C. A. West. 1969. *J. Biol. Chem.* **244**:3200.
Scott, A. I. 1970. *Accounts Chem. Res.* **3**:151. (Also see references cited therein).
Scott, A. I., and P. C. Cherry. 1969. *J. Amer. Chem. Soc.* **91**:5872.
Scott, A. I., G. T. Phillips, P. B. Reichardt, and J. G. Sweeny. 1970. *Chem. Commun.* p. 1396.
Scott, A. I., and C. C. Wei. 1972. *J. Amer. Chem. Soc.* **94** (in press).
Smith, G. F., R. T. Brown, J. S. Hill, K. Stapleford, J. Poisson, M. Muquet, and N. Kunesch. 1969. *Chem. Commun.* p. 1485.
Smith, G. N. 1968. *Chem. Commun.* p. 912.
Spenser, I. D. 1968. *In* "Comprehensive Biochemistry" (M. Florkin and E. H. Stotz, eds.), Vol. 20, Ch. 6. Elsevier, Amsterdam.
Walker, D. A., and A. R. Crofts. 1970. *Annu. Rev. Biochem.* **39**:389.
Ziegler, F. E., and E. B. Spitzner. 1970. *J. Amer. Chem. Soc.* **92**:3492.

BIOCHEMISTRY AND PHYSIOLOGY OF LOWER TERPENOIDS

W. DAVID LOOMIS and RODNEY CROTEAU

Department of Biochemistry and Biophysics, Oregon State University, Corvallis, Oregon

Introduction*

The term "lower terpenoids" as used here will refer to hemiterpenes, monoterpenes, and sesquiterpenes (C_5, C_{10}, and C_{15}). Monoterpenes and

* Abbreviations used throughout the text: MVA = mevalonic acid, IPP = isopentenyl pyrophosphate, DMAPP = dimethylallyl pyrophosphate, CoA = coenzyme A.

sesquiterpenes are major constituents of many essential oils, used as flavors, pharmaceuticals, perfumes, or solvents. The hemiterpene isoprene was only recently recognized as a natural product, but it was obtained very early in the history of organic chemistry by pyrolysis of rubber, and later by pyrolysis of oil of turpentine.

A brief chronology of the history of isoprene is illuminating, and sobering: first paper by Faraday on pyrolysis products of rubber, 1826; name "Faradayin" proposed for isoprene, 1835; name "isoprene" coined, 1860; structure elucidated, 1897 (Ruzicka, 1959). The chemistry of monoterpenes and isoprene occupied some of the ablest chemists for many years and forms the foundation of modern organic chemistry (Ruzicka, 1959). With this historical background it is not surprising that we now use the terms "terpenoid" (German, *Terpentin* = turpentine) and "isoprenoid" almost interchangeably in describing the whole class of compounds based on a branched C_5 unit related to isoprene or isopentane. The monoterpenes and sesquiterpenes are regarded as containing, respectively, 2 or 3 "isoprene" residues.

In contrast to isoprene and certain of the monoterpenes, the chemistry of sesquiterpenes could not be resolved adequately by classical methods and required the development of newer analytical techniques, especially gas chromatography-mass spectrometry (GC-MS) and infrared (IR) and nuclear magnetic resonance (NMR) spectroscopy. The number of known sesquiterpene structures has increased explosively in recent years, from one (farnesol) in 1913, to over 200 in 1964, to some 1000 in 1971 (Herout, 1971; Devon and Scott, 1972).

Modern analytical methods have also revolutionized the study of monoterpenes, of which some 400 are now known as natural products (Devon and Scott, 1972), and made possible the discovery of isoprene as a naturally occurring hemiterpene (Rasmussen, 1970; Sanadze and Dolidze, 1961).

Traditionally both mono- and sesquiterpenes have been thought of as components of essential oils, produced and accumulated or secreted in conspicuous amounts only by certain plants. It now seems probable that lower terpenoids are ubiquitous in higher plants. The classical essential oil producing plants probably are unique primarily in having specialized structures adapted to secreting or accumulating large quantities of these compounds, which in other plants are trace metabolites that either volatilize inconspicuously or are turned over metabolically.

Mono- and sesquiterpenes have also been thought of traditionally as volatile compounds, and the possible existence of nonvolatile derivatives such as glycosides or esters has, except in the case of the cyclopentanoid monoterpenes, been largely overlooked (Francis, 1971). For example, geraniol and nerol in rose petals were shown recently to occur largely as the β-D-glucosides (Francis and Allcock, 1969).

From the biochemical point of view, monoterpenes have attracted considerable attention, with relatively little return for the effort expended. Sesquiterpene biochemistry was neglected until recently, since even the chemistry of sesquiterpenes was inadequately understood. The biochemistry of hemiterpenes also was largely ignored, and isoprene, perhaps the most abundant natural hemiterpene, was regarded as not occurring in nature (Bonner, 1965; Robinson, 1967).

Several recent reviews have dealt with broad areas of the chemistry and biochemistry of monoterpenes (Loomis, 1967; Waller, 1969; Francis, 1971; Banthorpe *et al.*, 1972) and sesquiterpenes (Parker *et al.*, 1967; Herout, 1971) and with various specialized aspects of these compounds (for references, see Francis, 1971; Herout, 1971). There would seem to be no point in attempting to make the present review comprehensive and in so doing to re-review material that has already been treated elsewhere. For example, the rather specialized methylcyclopentanoid monoterpenes and their derivatives have been dealt with very competently (see Francis, 1971, or Banthorpe *et al.*, 1972, and numerous references therein) and will not be covered here at all. Rather we will attempt here to build a unified theory of *in vivo* biosynthesis of lower terpenoids in plants by integrating biochemical, physiological, and morphological observations, and at the same time to point out some of the possible pitfalls in studying this group of compounds. Thus, *in vivo* results will be emphasized, and *in vitro* studies will be discussed only to the extent that they aid in understanding the *in vivo* relationships. Our knowledge of the biochemistry and physiology of lower terpenoids has expanded considerably in recent years to a point at which *in vivo* and *in vitro* results both deserve individual attention. The *in vitro* studies will be reviewed separately elsewhere (Loomis and Croteau, submitted for publication).

Hemiterpenes

There are relatively few naturally occurring branched C_5 compounds. Figure 1 illustrates some examples. Although any compound with the appropriate C_5 skeleton may be regarded formally as isoprenoid, most of those found in nature probably are not biogenetically isoprenoid (i.e., derived from MVA). Valine certainly is not, and it is not generally regarded as isoprenoid in any sense. Isovaleric acid, dimethylacrylic acid, and several related aldehydes and alcohols probably arise from acyl-CoA esters

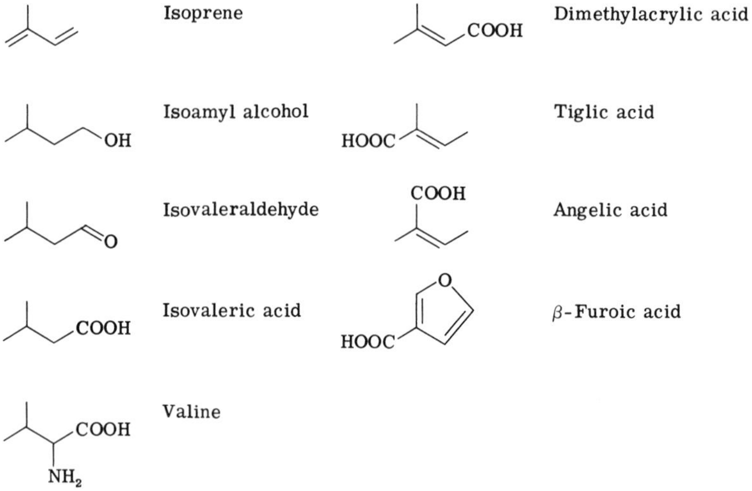

Isoprene

Dimethylacrylic acid

Isoamyl alcohol

Tiglic acid

Isovaleraldehyde

Angelic acid

Isovaleric acid

β-Furoic acid

Valine

Fig. 1. Some natural "hemiterpenoids."

formed as intermediates in the degradation of leucine, rather than via the isoprenoid pathway. Although terpene degradative pathways in higher plants are not known, microbial metabolism of certain terpenes (e.g., geraniol, farnesol) also yields dimethylacrylyl-CoA as an intermediate (Loomis, 1967), and dimethylacrylyl-CoA can be converted to β-hydroxy-β-methylglutaryl-CoA and then to MVA. In this sense, dimethylacrylyl-CoA and its derivatives are biogenetically related to isoprenoids, but also clearly distinct from them.

Isoprene

The most important natural hemiterpene may well be isoprene itself, long regarded as not occurring in nature. It really is not surprising that such a highly volatile and reactive compound should have gone undetected for so long, particularly in view of the long established notion that it did not exist as a biological product. The case of isoprene is an example, however, of something that has been too common in the area of the lower terpenoids: the danger of confusing preconceived notions with facts. Isoprene has recently been detected as a natural emission from leaves of a number of woody plant species (Rasmussen, 1970; Sanadze and Dolidze, 1961). Rasmussen (1970) confirmed the identity of isoprene from five tree species, representing five families, by gas chromatography and infrared and mass spectrometry.

Emission of isoprene from leaves occurs only during and for a few minutes after illumination (Sanadze and Kursanov, 1966; Rasmussen and Jones, 1972). Results from feeding of ^{14}C-labeled compounds to illuminated leaves of *Populus nigra* (Sanadze, 1966) suggested that isoprene is preferentially formed from products of photosynthesis. $^{14}CO_2$ was the best precursor; glucose-1-^{14}C was utilized 2.5–3 times as well as glucose-6-^{14}C; and acetate-1-^{14}C was utilized 2- to 2.5-fold better than acetate-2-^{14}C. Labeled MVA was not tested. Recently, Sanadze *et al.* (1972) have extended these studies, using $^{13}CO_2$.

Although photosynthetically fixed CO_2 appeared to be the preferred substrate for isoprene biosynthesis, high levels of CO_2 inhibited the emission of isoprene (Sanadze, 1964; Rasmussen and Jones, 1972). The critical CO_2 level could not be determined accurately but appeared not to be far above the normal atmospheric level. *Populus* leaves illuminated in sealed chambers with an initial CO_2 concentration of 1.3%, fixed CO_2 steadily but did not produce isoprene for several hours, until the CO_2 concentration was reduced to about 0.03% or less (Sanadze, 1964).

All these observations tempt one to suggest that isoprene is synthesized in the chloroplasts, from a photosynthetically produced precursor, very possibly glycolate. Glycolate is a principal product of photosynthesis in many plants at normal atmospheric CO_2 concentrations, while its formation is inhibited by higher levels of CO_2 (Zelitch, 1965, 1969; Jackson and Volk, 1970). Glycolate commonly serves as a substrate of photorespiration, and photorespiration is also inhibited by high concentrations of CO_2 (Jackson and Volk, 1970).

Rogers, Shah, and Goodwin (1968) have shown that exogenous glycolate-^{14}C, as well as glycine-^{14}C or serine-^{14}C (also intermediates of photorespiration), can be utilized for the biosynthesis of chloroplastidic terpenoids (e.g., β-carotene) that are otherwise synthesized preferentially from photosynthetically fixed CO_2. From this and related evidence they propose that acetyl-CoA (and hence MVA) for terpenoid biosynthesis within the chloroplasts can arise from photosynthetically formed glycolic acid, via glycine and serine.

It must be recognized that all the isoprene studies have measured emission of isoprene from leaves or leaf disks, not biosynthesis per se. Emission of a gas from the leaves requires not only that the substance be produced, but that the stomates be open, allowing its release. High CO_2 concentrations, such as those that inhibit glycolate production, photorespiration, and isoprene emission, also cause closure of stomates (Zelitch, 1969). Rasmussen and Jones (1972) stated that attempts to correlate observations of isoprene emission with stomatal activity were "inconclusive."

However, evidence from their light-dark emission measurements suggests that isoprene biosynthesis per se is coupled to photosynthesis. If the non-emission of isoprene in the dark, in such experiments were due primarily to closure of the stomates, one would expect isoprene to build up inside the leaves in the dark and to be released in a burst when the stomates reopen in the light. Instead of this, the rate of isoprene emission was observed to increase slowly, after an initial lag of about 2 minutes, after the lights were turned on (Rasmussen and Jones, 1972).

Thus it appears that isoprene is an important plant product, whose production is closely linked to photosynthesis. The biosynthesis of isoprene may represent an alternative to, or a variant of, photorespiration (Jackson and Volk, 1970). Further developments in this area should be of great interest.

Mono- and Sesquiterpenes

Earlier biochemical studies in this area dealt primarily with monoterpenes, with little consideration given to sesquiterpenes. Recent studies have dealt with the two groups simultaneously, and it is convenient to treat them together here. Figure 2 shows the structures of several monoterpenes which will be discussed. These are typical monoterpenes but by no means represent the great diversity of structures that exist in nature. Pertinent sesquiterpene structures will be illustrated later, in connection with biosynthetic studies.

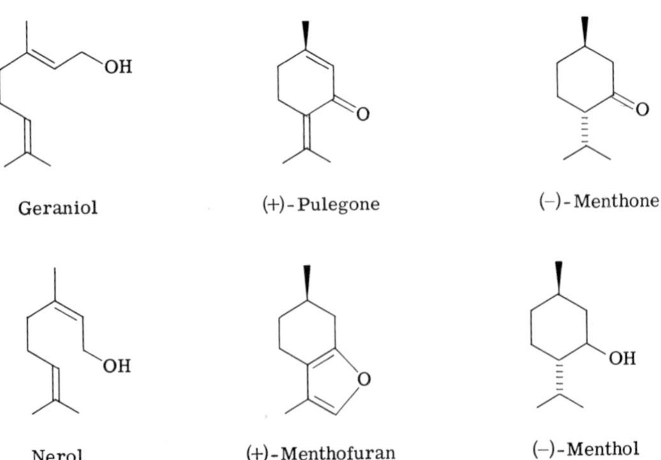

FIG. 2. Selected monoterpene structures.

INCORPORATION OF LABELED PRECURSORS

One of the most striking findings to emerge from the numerous *in vivo* tracer studies of monoterpene biosynthesis is the almost universally poor incorporation of exogenous labeled substrates, especially MVA-^{14}C. Typically from 0.01 to 0.1 percent of MVA or acetate label is incorporated into monoterpenes (Loomis, 1967; Francis, 1971; Banthorpe *et al.*, 1972). The only notable exception to date is the case of rose petals (Francis and O'Connell, 1969), which will be discussed below. In some instances, slightly higher incorporations of MVA label into "volatile oils" have been reported, but without positive identification of the labeled components (e.g., Nicholas, 1962; Hefendehl *et al.*, 1967). Poor incorporations of MVA-^{14}C into certain sesquiterpenes were also reported (Biollaz and Arigoni, 1969; Corbella *et al.*, 1969), but these studies were much less extensive, and few experimental details were published. In several cases (Biollaz and Arigoni, 1969; Corbella *et al.*, 1969; Banthorpe and Baxendale, 1970; Banthorpe *et al.*, 1970; Croteau and Loomis, 1972a) degradation of the mono- and sesquiterpene products indicated direct incorporation of MVA-2-^{14}C via the isoprenoid pathway, but in other cases extensive randomization of label was observed (see Loomis, 1967; Francis, 1971, Banthorpe *et al.*, 1972).

Such results might suggest an alternate, nonmevalonoid, pathway for the biosynthesis of mono- and sesquiterpenes were it not for the degradative evidence cited, and the absence of other data supporting alternative pathways. Low incorporations might also be attributed to poor uptake of MVA by the plant or to the inhibition of essential enzymes by the concentrated doses of MVA often administered. However, a number of investigators have carried out translocation and substrate-dilution studies, and even when optimum dose levels and methods of administration are utilized, high levels of MVA incorporation (i.e., greater than 1%) are not realized (Regnier *et al.*, 1968; Horodysky *et al.*, 1969; Banthorpe *et al.*, 1970; Croteau and Loomis, 1972a).

The "physiological" source of carbon in green plants is CO_2, and $^{14}CO_2$, in the light, was found to be a relatively good monoterpene precursor, much better than MVA-^{14}C or acetate-^{14}C (Loomis, 1967; Burbott and Loomis, 1969; Hefendehl *et al.*, 1967). From these results we suggested (Loomis, 1967) that the site of monoterpene biosynthesis was isolated from the rest of the plant and that the bulk of MVA utilized in monoterpene synthesis must arise at the site of synthesis from translocated photosynthate, probably sugars.

In search of other substrates that might be able to penetrate to the apparently isolated site of synthesis, we tested a number of labeled compounds as monoterpene precursors, using as an experimental system stem-

TABLE 1

INCORPORATION OF ^{14}C-SUBSTRATES INTO PEPPERMINT MONOTERPENES[a]

^{14}C-Substrate	Percent incorporation
Acetate-2-	ND[b]
Mevalonate-2-	0.01
Glycolate-2-	0.04
Glyoxylate-2-	0.04
Pyruvate-2-	0.02
Serine-3-	0.05
Alanine-2-	0.07
Ribose-1-	0.07
Glucose-1-	0.09
Glucose-2-	0.16
Glucose-3,4-	0.14
Glucose-6-	0.38
CO_2 (6 hours)	0.2
CO_2 (34 hours)	1.9

[a] A 6-hour incorporation period in the light, in a growth chamber, was used in each experiment, except where otherwise noted. $^{14}CO_2$ was fed at a level of about 1 μmole per cutting, with 1 hour exposure time. All other substrates were fed at a level of 3 μmoles per cutting. Mevalonate and serine were fed at a level of 3 μmoles of RS-form, but percent incorporation was calculated on the basis of the physiological isomer only (Burbott and Loomis, unpublished data).

[b] ND = not detectable.

fed vegetative cuttings from peppermint plants grown under optimum conditions for terpene production.* The results of one such experiment, in which the products were detected by gas radiochromatography, are shown in Table 1. Glucose and CO_2 were the most efficient substrates, and acetate and mevalonate were the least effective. Glycine-2-^{14}C was also

* In general in our work, as described here, peppermint cuttings consist of the growing tip plus two leaf pairs and weigh about 300 mg each. Plants are grown in a growth chamber under 16 hour-day and cool night conditions as described by Burbott and Loomis (1967). Cuttings are taken from 2 to 4 hours after the beginning of the light period unless otherwise specified and are matched as closely as possible visually. In spite of these precautions, some variability is unavoidable, especially when comparing separate experiments that may have been done several months apart.

tested in an earlier experiment in which the products were counted on thin-layer plates. The incorporation of glycine label was 0.01–0.02 percent in 6 hours. The high incorporation of glucose, and especially of glucose-6-^{14}C, suggested a preferential transport of sugars. Thus, the findings would appear to support our earlier suggestion that the site of monoterpene biosynthesis in peppermint is compartmentalized, and isolated from the mainstream of plant metabolism.

Although MVA-2-^{14}C label was not appreciably incorporated into monoterpenes when fed to peppermint cuttings, gas radiochromatography of hexane extracts of the cuttings revealed label in compounds with longer retention times, corresponding to sesquiterpenes. This higher molecular weight fraction was then analyzed by GC-MS and shown to contain a series of sesquiterpene hydrocarbons, including caryophyllene, bourbonenes, cadinenes, and muurolenes. Caryophyllene and γ-muurolene were the major sesquiterpenes, and, under the growth conditions employed in this study, caryophyllene comprised over 50 percent of the sesquiterpene fraction and up to 0.8 percent of the essential oil extracted from the cuttings.

A time-course study of the incorporation of MVA-2-^{14}C into mono- and

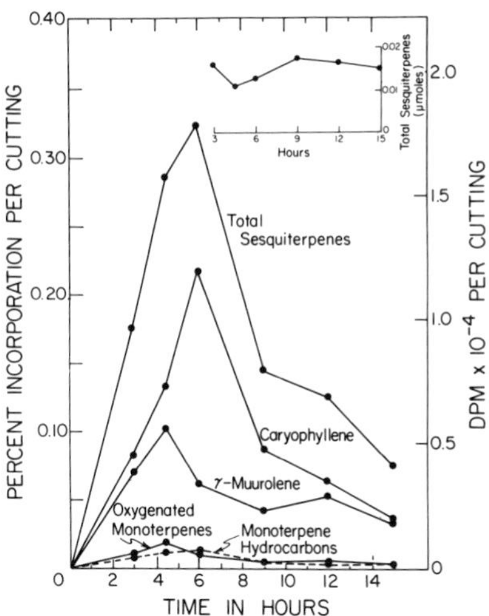

FIG. 3. Time course of labeling of peppermint mono- and sesquiterpenes from mevalonic acid-2-^{14}C. Data presented are from a single set of matched cuttings fed 1 μmole of *RS*-MVA-2-^{14}C per cutting. Data points for total sesquiterpenes and monoterpenes represent the sums of activities in detectable gas radiochromatographic peaks. Time is from start of MVA feeding, and percent incorporation is calculated by assuming that only *R*-MVA is physiologically active. From Croteau and Loomis (1972a).

sesquiterpenes of peppermint cuttings was then carried out; the results are shown in Fig. 3. The most striking aspect of this time-course study was the fact that approximately 90 percent of the MVA label incorporated into the essential oil was in the sesquiterpenes, which comprise less than 2 percent of the oil. Caryophyllene incorporated label to the greatest extent, followed by γ-muurolene which acquired maximum label somewhat earlier than caryophyllene. All the other sesquiterpenes incorporated some label, and these activities are included in the total sesquiterpene plot. Since the monoterpenes acquired very little label they are plotted as classes (hydrocarbons vs. oxygenated) rather than as individual components.

We then repeated our earlier studies using ^{14}C-labeled sugars and $^{14}CO_2$ as precursors, and followed the incorporation of label into sesquiterpenes as well as monoterpenes. A time-course of glucose-U-^{14}C incorporation is shown in Fig. 4. In this instance the monoterpenes incorporated considerably more label than the sesquiterpenes (1.17 percent vs. 0.06 percent incorporation), in a ratio roughly approximating the natural proportion of mono- and sesquiterpenes in the plant. Similar results were obtained when sucrose-U-^{14}C or $^{14}CO_2$ was administered to peppermint cuttings in the light (see Fig. 5). The evidence suggests to us the presence of separate

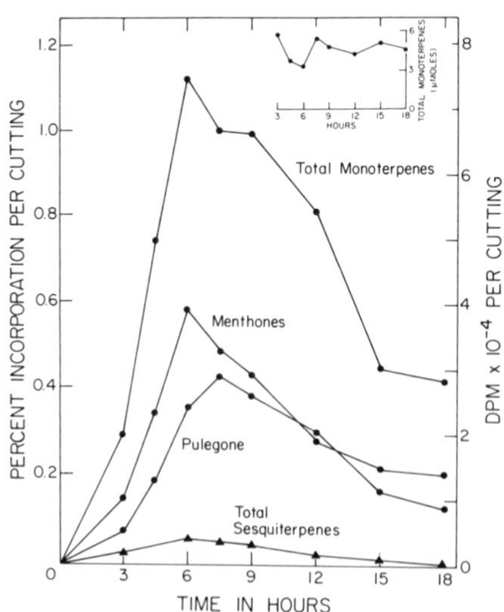

FIG. 4. Time course of labeling of peppermint mono- and sesquiterpenes from glucose-U-^{14}C. Data presented are from a single set of matched cuttings fed 1 μmole of glucose-U-^{14}C per cutting. Data points for total mono- and sesquiterpenes represent the sums of activities in detectable gas radiochromatographic peaks. Time is from start of glucose feeding. From Croteau *et al.* (1972a).

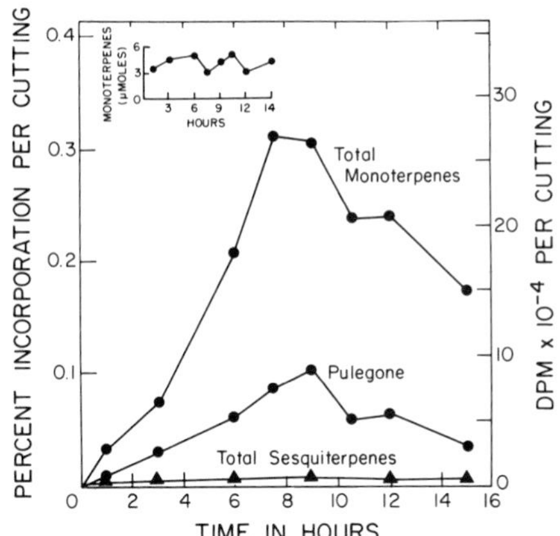

FIG. 5. Time course of labeling of peppermint mono- and sesquiterpenes from $^{14}CO_2$ in continuous light. Data presented are from a single set of matched cuttings fed 0.71 μmole $^{14}CO_2$ per cutting. Data points for total mono- and sesquiterpenes represent the sums of activities in detectable gas radiochromatographic peaks. Time is from start of $^{14}CO_2$ exposure. From Croteau *et al.* (1972a).

sites for monoterpene and sesquiterpene biosynthesis, the sesquiterpene site being the more accessible to exogenous MVA-2-^{14}C, while the monoterpene site is more accessible (or at least as accessible) to exogenous glucose-^{14}C, sucrose-^{14}C, and photosynthetically fixed $^{14}CO_2$. We have also noted a similar pattern of differential labeling between mono- and sesquiterpenes in preliminary studies with hop (*Humulus lupulus* L.) cuttings fed MVA-2-^{14}C and glucose-U-^{14}C, which suggests that this phenomenon may be widespread in plants.

The mono- and sesquiterpenes have traditionally been found together in essential oil extracts and distillates, and there has been a natural tendency to regard these two types of compounds as closely related, both with regard to biochemical precursors and to site of synthesis. This assumption is not supported by experimental data, and the present results suggest that it is incorrect. These results also emphasize the danger of assuming, for example, that labeled "volatile terpenes" are monoterpenes, or even terpenoid, without careful characterization. Similarly, it should not be assumed that label from MVA incorporated into "terpene hydrocarbon fractions" of hexane extracts is necessarily in "carotenes" or "lower terpenes" because it is likely to be mainly in squalene (see below).

Although the incorporation of MVA-^{14}C into sesquiterpenes was considerably greater than into monoterpenes of peppermint (maximum = 0.33 percent of *R*-form vs. 0.03 percent in 6 hours), the combined total incorporation into both groups of compounds was still only 0.36 percent in 6 hours. This compares with 22 percent incorporation of *R*-MVA into rose petal monoterpenes in 1 hour (Francis and O'Connell, 1969). This raises an obvious question as to whether exogenous MVA can be efficiently utilized in the biosynthesis of any isoprenoid compound in peppermint. To answer this question we studied the incorporation of MVA-2-^{14}C into the triterpene squalene and into sterols, substances that are ubiquitous in higher plants. Earlier work by Nicholas (1962) on the labiate *Ocimum basilicum* suggested that MVA-2-^{14}C was an efficient precursor of squalene and other triterpenes although it was not a very efficient precursor of steam volatile substances.

Figure 6 shows a time course of MVA-2-^{14}C incorporation into squalene and sterols of peppermint cuttings. The incorporation of MVA into these compounds was very much greater than into lower terpenes, although there is less squalene (0.01 μmole) and sterols (0.05 μmole) per cutting than monoterpenes (3 μmole) or sesquiterpenes (0.05 μmole). A further observation from this study was that most of the radioactivity in the hydrocarbon fraction isolated from MVA-^{14}C-fed cuttings (containing mono- and sesquiterpene hydrocarbons, squalene, and carotenoids) was in squalene. The evidence suggests that triterpenes, carotenoids, and lower terpenes are synthesized at different sites, with the carotenoid and lower terpene sites being subject to a greater degree of compartmentalization than the squalene–triterpene sites.

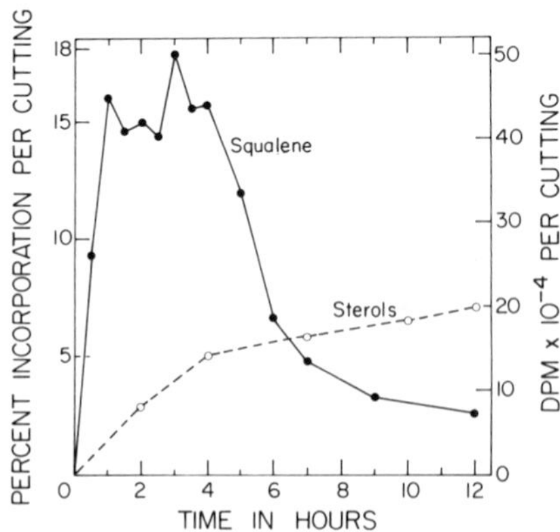

Fig. 6. Time course of labeling of squalene and sterols in peppermint from mevalonic acid-2-^{14}C. Data presented are from a single set of matched cuttings fed 0.5 μmole *RS*-MVA-2-^{14}C per cutting. Squalene was determined gas radiochromatographically, and sterols via scintillation counting of a thin-layer chromatography isolate. Time is from start of MVA feeding, and percent incorporation is calculated by assuming that only *R*-MVA is physiologically active. From Croteau and Loomis (1972b).

It will be noted that our data are presented as percent incorporation rather than as specific activity. Specific activities are meaningful only if they truly represent metabolic pools. The presence of large and variable storage pools of many terpenes and other plant products means that specific activities, as measured, often do not represent metabolic pools. Failure to appreciate this can result in completely erroneous interpretation of tracer data. There is no really satisfactory way of expressing the results of tracer incorporation experiments in plants. "Percent incorporation" has the disadvantage that values vary with the amount of substrate fed. This is illustrated by comparison of 6-hour MVA-^{14}C incorporation in Table 1 with Fig. 3 (i.e., decreasing the quantity of MVA fed from 3 μmoles to 1 μmole increased the percent incorporation from 0.01 to 0.03 percent). Percent incorporation does provide a reasonable compromise and yields a simple and valid means of comparing data when substrate levels are known.

The usefulness of time-course experiments should also be stressed. Kinetic analysis of the absolute amounts of label in individual components can provide evidence for biosynthetic sequences, even when there are gross differences in pools and specific activities.

Dynamic Metabolism of Mono- and Sesquiterpenes

Tracer Studies

Earlier (Loomis, 1967; Burbott and Loomis, 1969) we reviewed several kinds of evidence for metabolic turnover of monoterpenes in plants. In particular, we presented data from $^{14}CO_2$ time courses with peppermint in which monoterpenes became labeled and then in the course of a few hours lost most of their label. Details of the time courses varied, depending on the conditions, but the turnover was consistently observed. Evidence was presented to indicate that the observed turnover was not due simply to evaporation. Similar $^{14}CO_2$ time-course experiments with peppermint by Hefendehl and co-workers (1967) also suggested turnover of monoterpenes. Time-course studies with peppermint cuttings using glucose-^{14}C and sucrose-^{14}C have now indicated that monoterpenes derived from these precursors undergo similar turnover (compare Fig. 4 and Fig. 5). Turnover of monoterpenes derived from exogenous MVA-2-^{14}C has been suggested by data from a number of studies (Scora and Mann, 1967; Banthorpe *et al.*, 1970; Croteau and Loomis, 1972a) with several plant species, but the generally low incorporation of MVA into these compounds has prevented an accurate assessment (for example, see Fig. 3). In the singular case of rose petals, however, MVA is readily incorporated into both free monoterpenes and monoterpene glucosides (up to 22 percent incorporation of the *R*-isomer in 1 hour) and, in this instance, the label is almost completely lost after 4 hours (Francis and O'Connell, 1969). In the case of rose petals, evaporation may be a factor, but if it is, the labeled monoterpenes must be evaporating preferentially at a time when the total monoterpene content of the petals is increasing (Francis and Allcock, 1969).

Time-course studies using MVA-2-^{14}C as a terpene precursor in peppermint cuttings have now indicated that sesquiterpenes in this plant also undergo metabolic turnover (Fig. 3). This evidence appears to be the first demonstration of turnover of sesquiterpenes in plants and adds to the growing body of evidence indicating that lower terpenoids and other "secondary plant products" are metabolically active and are not biosynthetic "dead ends."

Figures 3 and 4 also show the mono- and sesquiterpene content per cutting throughout the time-course experiments with peppermint. In each case the cutting-to-cutting variation in micromoles exceeds the amount of ^{14}C-labeled precursor incorporated and may represent differences in essential oil storage pools of the cuttings. Within the rather narrow limits observed in our experiments (i.e., a 2-fold variation) the apparent pool size bore little relation to the ability of the cutting to utilize exogenous precursors. Storage pools, such as the oil contained in glandular secretory spaces, probably turn over quite slowly, and the turnover of mono- and sesquiterpenes observed in these experiments most likely represents the turnover of a more active "metabolic" pool. As present techniques cannot readily distinguish between these pools, turnover data necessarily represent the overall changes occurring in the cuttings. It might be mentioned again at this point, that because of variations in internal pool sizes, time-course curves on a percent incorporation basis are more reproducible than time-course curves on a specific activity basis.

Most short-term time-course studies thus far carried out have indicated that mono- and sesquiterpenes biosynthesized from labeled precursors undergo considerable metabolic turnover within a few hours. Time-course studies based on long incorporation periods (i.e., time scale in days rather than hours) are likely to overlook this initial activity. Furthermore, studies with peppermint and other species indicate that administered precursors are rapidly utilized by the plant in a variety of pathways. For example, when peppermint cuttings were fed 1 μmole *RS*-MVA-2-^{14}C in 0.1 ml water through the cut stem, 1–1.5 hours were required for complete uptake. Two hours after the beginning of feeding, 25 percent of the total activity was easily extracted in hexane, while somewhat less than 50 percent of the total activity (presumably representing primarily the nonphysiological

S-MVA) was water extractable (Croteau and Loomis, unpublished observations). Thus, under these conditions, a significant proportion of *R*-MVA-2-^{14}C appears to be utilized by the plant almost as rapidly as it is taken up and translocated, although only a very small proportion of this label (<1 percent) is incorporated into lower terpenes. Similarly, ^{14}C-labeled terpinen-4-ol fed to *Tanacetum* was converted within a 44-hour period into monoterpenes, 9 percent; carotenoids, 3.5 percent; chlorophyll, 9 percent; amino acids, 4.0 percent; sugars, 7 percent; and tissue-bound substances, 3.5 percent (Banthorpe and Wirz-Justice, 1969). The remaining activity was widely distributed in unidentified, but primarily water-soluble, materials. Long-term time course studies are thus not likely to reflect either the incorporation of the original ^{14}C-labeled precursor into lower terpenes or the simple turnover of these compounds, but rather a very complex interaction of many synthetic and degradative pathways.

It can be argued that the apparent turnover observed in time-course studies using exogenous precursors represents "induced" metabolism caused by the presence of excess substrate and, as such, is nonphysiological and does not reflect true *in vivo* metabolic turnover. To a certain degree this is true. Cuttings and intact plants exhibit somewhat different turnover phenomena. [Experiments have shown that on exposure to $^{14}CO_2$ peppermint cuttings incorporate somewhat more label into mono- and sesquiterpenes and show a faster rate of turnover than intact plants (Croteau *et al.*, 1972b).] Also, as one might expect, the observed turnover is very dependent on environmental conditions (e.g., light and temperature) and on the physiological condition of the plant. For example, when peppermint cuttings are taken from plants at the end of a normal 16-hour daylight period and allowed to take up and metabolize MVA-^{14}C in the light, they incorporate more MVA-^{14}C label into mono- and sesquiterpenes at the 6-hour peak period than do cuttings taken 2 hours after the start of the light period, and the turnover period is greatly extended (Croteau *et al.*, 1972b). Finally, the observed turnover is dependent on the amount of precursor fed, the method of administration, and the utilization of precursor in other pathways (i.e., residence time of the precursor as influenced by total synthesis, degradation, and resynthesis taking place).

In spite of the limitations and variability of time-course experiments, the observed turnover of labeled terpenes does indicate that the biosynthetic system of the plant is capable of rapidly metabolizing mono- and sesquiterpenes and that most of the labeled terpene produced in short-term experiments is indeed metabolized and is not stored. The variability of time-course and turnover behavior in response to environmental and physiological effects should not be regarded as a weakness of this type of data, but rather as a means of understanding mono- and sesquiterpene biosynthesis and metabolism and of integrating these pathways into the overall physiology of the plant. For example, variation of turnover period with the time of day that cuttings were taken suggested that terpene biosynthesis is dependent on the amount of endogenous photosynthetate available, and that terpene storage in particular is enhanced by an abundance of photosynthate.

Periodic Analyses

Further evidence for metabolic turnover of lower terpenes has been obtained from periodic analyses of essential oil-producing plants. A number of investigators, notably Flück and his co-workers (see Weiss and Flück, 1970, and citations therein; also Schröder, 1969), have made extensive analyses to test for possible diurnal fluctuations of essential oils in plants. In several species the analyses indicated an increase of essential oil during the day, reaching a peak at some time in the afternoon or evening, and a decrease during the night. Weiss and Flück (1970) pointed out that the increase in essential oil during the heat of the day is contrary to what one would expect, based on anticipated increased evaporation at higher temperatures. They suggested that enhanced terpene synthesis during the day more than compensates for evaporative losses. Banthorpe and Wirz-Justice (1969) reported that the yield of essential oil from *Tanacetum vulgare* was about 20% greater at night than by day. Since the hours of sampling were not indicated, it is not possible to compare these results with those of Flück and others. Banthorpe and Wirz-Justice, however, did show that the observed variation in yield was not due to evaporative losses.

Evidence for monoterpene turnover on a time scale of weeks was obtained by periodic leaf-by-leaf analyses of peppermint grown in the growth chamber (Burbott and Loomis, 1969, and unpublished observations). Two growing conditions were compared: "cool-night", with 16-hour day at 25°C and 8-hour night at 8°C; and "warm-night," with 14-hour day at 25°C and 10-hour night at 25°C. The total monoterpene content for the ninth leaf pair from the base (representative of the larger mid-stem leaves) is shown in Fig. 7. The cool-night plants accumulated large amounts of monoterpenes until the time that floral initiation could first be recognized macroscopically (A), and then lost much of the oil. Warm-night plants behaved quite differently. The leaves accumulated monoterpenes more slowly and reached a constant level, which did not change greatly after blooming. This constant level corresponded to the postbloom level in the cool-night plants. A possible explanation is that the amount of essential oil seen in the warm-night plants represents filling of relatively inert

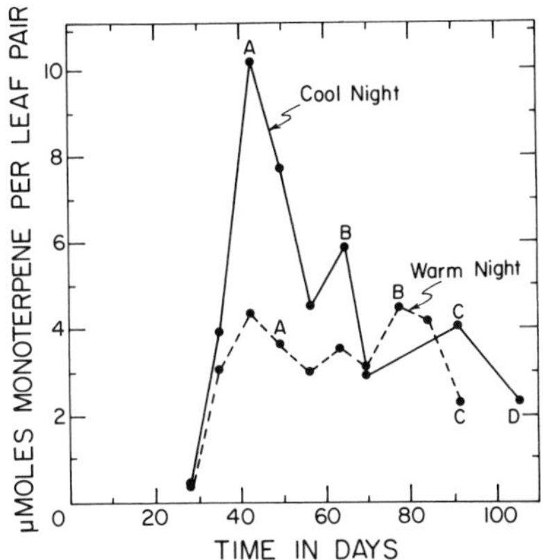

FIG. 7. Periodic analysis of total monoterpenes of leaf pair number 9 from peppermint plants grown under controlled environmental conditions. "Cool night": 16-hour day, 25°C; night, 8°C. "Warm night": 14-hour day, 25°C; night, 25°C. A = Time at which floral initiation could be recognized macroscopically. B = time at which first flowers opened; C = full bloom; D = end of bloom. From Burbott and Loomis (unpublished observations).

storage pools, while the excess seen in the cool-night plants represents accumulation at or near active metabolic pools as a result of greater availability of photosynthate under these conditions. That there was more available photosynthate in the cool-night plants was indicated by enhanced formation of anthocyanin pigment in these plants. When photosynthate is later diverted for flower production, these leaves would lose their "energy-rich" status and metabolize those monoterpenes that are in or near the metabolic pools. Finally, only the terpenes in the storage pools would remain. However, the storage pools apparently are also subject to slow metabolism, as menthols replace the menthones as the plant ages (Burbott and Loomis, 1969).

As we have seen, there is a variety of evidence suggesting metabolic turnover of mono- and sesquiterpenes in plants. The evidence suggests further, that synthesis, turnover, and storage of essential oils are controlled by the balance between photosynthesis and utilization of photosynthate. Catabolism of essential oil components during times of photosynthate deficiency does not seem unreasonable, as the mono- and sesquiterpenes represent a considerable amount of potential metabolic energy. For example, Francis (1971) has calculated that oxidation of menthone would provide an ATP yield intermediate between glucose and fatty acids, if a pathway similar to the microbial oxidation of geraniol were utilized.

Nonequivalent Labeling of C_5 Units

One of the most unexpected features of the biosynthesis of lower terpenes was revealed by chemical degradation of mono- and sesquiterpenes derived from exogenous ^{14}C-labeled precursors. The mono- and sesquiterpenes are thought to originate via the "classical" condensation of a DMAPP starter with, respectively, one or two IPP units. As both IPP and DMAPP are formed *in vivo* from MVA, and IPP is the immediate precursor of DMAPP, the biosynthesis of mono- and sesquiterpenes from exogenous ^{14}C-labeled precursors, such as MVA-2-^{14}C or $^{14}CO_2$, would be expected *a priori* to result in equivalent amounts of ^{14}C-labeled tracer being incorporated into IPP- and DMAPP-derived moieties of the terpene molecules. Although such equivalency of labeling was observed in certain monoterpenes synthesized by flowers (Francis *et al.*, 1970; Godin *et al.*, 1963), the biosynthesis of mono- and sesquiterpenes in other tissues from several species has yielded preferentially labeled terpenes, containing the bulk of incorporated tracer in the IPP-derived portion of the molecule.

Banthorpe and co-workers (1970) found that MVA-2-^{14}C was specifically incorporated into thujone, isothujone, and sabinene in cuttings from *Thuja*, *Tanacetum*, and *Juniperus* but that from 90 percent to over 99 percent of the incorporated tracer was located in the IPP-derived portions of the molecules. Camphor (Banthorpe and Baxendale, 1970), α-pinene and pulegone (Banthorpe *et al.*, 1972), and artemisia ketone (Banthorpe and Charlwood, 1971) derived from MVA-2-^{14}C in several species were also shown to be preferentially labeled in the IPP-derived moiety. In these experiments, the MVA-2-^{14}C was stem fed over a 4- to 5-day period with incorporations of about 0.02 percent. That this type of preferential labeling is not restricted to the monoterpenes was shown by degradation of the picrotoxane sesquiterpenes coriamyrtin and tutin derived from MVA-2-^{14}C in *Coriaria japonica*, in which 80 percent of the incorporated radioactivity of both compounds was found to reside in the IPP portion of the molecule, as opposed to the theoretical 66.7 percent (Biollaz and Arigoni, 1969). Similarly, caryophyllene derived from MVA-2-^{14}C in peppermint cuttings after a 6-hr incorporation period contained 88 percent of the incorporated label in IPP-derived carbons compared to the theoretical 66.7 percent (see Fig. 8).

The poor incorporation of MVA-^{14}C and acetate-^{14}C into lower terpenoids, especially monoterpenes, has made degradative studies very difficult. At the same time, it was taken for granted that $^{14}CO_2$ or ^{14}C-sugars, which were incorporated much more effectively, would yield randomly labeled products. The studies just described suggested that this might not be so, and prompted degradative studies of monoterpenes biosynthesized from $^{14}CO_2$. Wuu and Baisted (1972) degraded the acyclic monoterpene geraniol

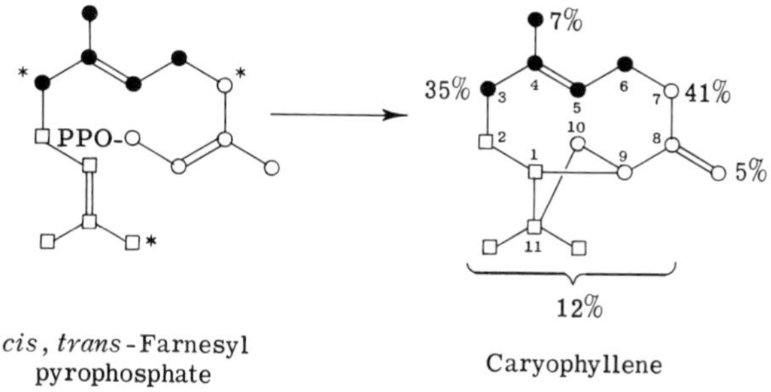

cis, trans-Farnesyl pyrophosphate

Caryophyllene

FIG. 8. Hypothetical biogenesis of caryophyllene via *cis,trans*-farnesyl pyrophosphate derived from mevalonic acid-2-^{14}C, and experimentally determined distribution of label in caryophyllene derived from MVA-2-^{14}C. Labeled carbon atoms of farnesyl pyrophosphate theoretically derived from MVA-2-^{14}C (each containing $33\frac{1}{3}\%$ of incorporated label) are identified by asterisks. Open squares represent carbon atoms derived from DMAPP. Filled circles represent carbon atoms derived from the first IPP added, and open circles represent carbon atoms derived from the second or terminal IPP. Bracket indicates the portion of the caryophyllene molecule containing carbons 1, 2, 8, 9, 10, 11, and 11,11-dimethyl. From Croteau and Loomis (1972a).

(Fig. 2), from *Pelargonium graveolens* (rose geranium), while Croteau *et al.*, (1972a) degraded the cyclic monoterpene pulegone (Fig. 2), from peppermint. In both cases oxidative cleavage provided a comparison of label in the isopropylidene group, derived hypothetically from DMAPP, with the total label. Calculated label in the DMAPP-derived moiety was obtained by multiplying the isopropylidene label by $\frac{5}{3}$. In pulegone derived from $^{14}CO_2$ in peppermint, degradations at 1, 3, 6, 9, and 12 hours after $^{14}CO_2$ exposure (see time course in Fig. 5) indicated that virtually all the label was in the IPP-derived half of the molecule.

The results of degradations of geraniol from rose geranium were more complex. In shorter time periods, when very small amounts of labeled geraniol had been synthesized (see Fig. 9), the label appeared to be almost equally distributed between the two C_5 units. At 12 hours, with much higher levels of labeled geraniol, 78 percent of the incorporated label was in the IPP-derived half of the molecule. At 24 hours (after 75 percent of the labeled geraniol had turned over), only 59 percent of the geraniol label was in the IPP-derived moiety. Thus, within 1 day IPP and DMAPP moieties had almost equilibrated in geraniol from rose geranium.

Several investigators (Banthorpe *et al.*, 1972; Croteau and Loomis, 1972a; Wuu and Baisted, 1972) have suggested that preferential labeling may result from the condensation of IPP derived from exogenous ^{14}C-labeled precursor with DMAPP that is mainly present in a metabolic pool. Other explanations, such as a nonmevalonoid origin of DMAPP,

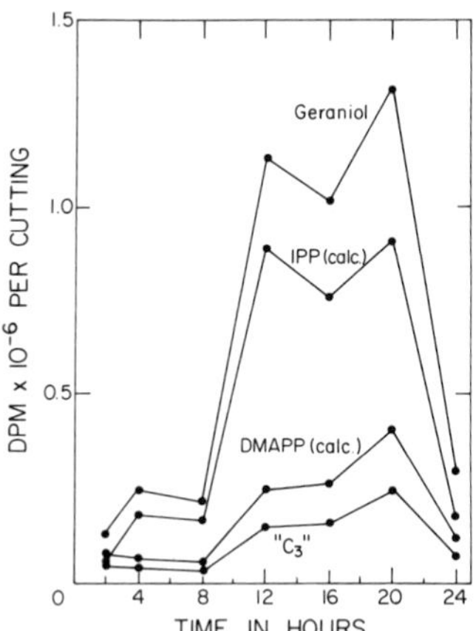

FIG. 9. Time-course of labeling of geraniol in rose geranium from $^{14}CO_2$. Cuttings were exposed to $^{14}CO_2$ for 2 hours followed by 22 hours metabolism in circulating air in the light. "C_3" refers to the measured radioactivity in the geraniol isopropylidene group. Label in DMAPP- and IPP-derived moieties are calculated. From Wuu and Baisted (1972).

might also explain the observed labeling patterns, but they seem less likely. We have previously suggested that the compartmentalization of mono- and sesquiterpene biosynthesis, and the apparent participation of a DMAPP pool, may be related phenomena. Certainly, compartmentalization of metabolic intermediates with their related enzymes is a well documented phenomenon in plants (Oaks and Bidwell, 1970), and the concept of an endogenous DMAPP pool is consistent with the equilibrium ratio of IPP to DMAPP conversion (about 1:10) as established by *in vitro* studies of IPP-isomerase (Agranoff *et al.*, 1960; Shah *et al.*, 1965; Holloway and Popják, 1967; Green, 1971) and with the suggestion from studies on germinating peas that the IPP-DMAPP condensing enzyme (i.e., prenyltransferase) is rate-limiting in isoprenoid biosynthesis (Green, 1971; Green and Baisted, 1971). If the biosynthetic sites for mono- and sesquiterpenes in most tissues are energy deficient, as we have proposed (see following section on Compartmentalization of Biosynthetic Sites), such an accumulation of an energetically favored metabolic intermediate does not seem unlikely.

In contrast to the mono- and sesquiterpenes, squalene and sterols of peppermint readily incorporated exogenous MVA-2-^{14}C (compare Figs. 3 and 6). The squalene-^{14}C synthesized was degraded at several periods during the time course (Fig. 6) and shown to be equivalently labeled, according to theory, in the IPP- and DMAPP-derived units (Croteau and

Loomis, 1972b). Thus, the bulk of triterpenoid biosynthesis does not appear to be tightly compartmentalized in peppermint, and a DMAPP pool does not appear to participate in the biosynthesis of these compounds. Because of the relatively rapid rate of synthesis of triterpenes, however, an endogenous DMAPP pool, if present, might be rapidly diluted with MVA-2-^{14}C label. Thus, the effects of a pool may only be evident when the amount of incorporation of labeled precursor is low enough, or the pool is large enough, to allow a detectable proportion of the DMAPP moieties to remain unlabeled.

The results in peppermint suggest rapid synthesis and turnover of terpenoids at the squalene-triterpene sites, with no apparent DMAPP pool, contrasted with slower synthesis, and a DMAPP pool, at the "essential oil sites." Based on this concept, a possible explanation of Wuu and Baisted's results (Fig. 9) might be proposed. It is possible that two distinct geraniol pools were being examined in these studies: the early, almost uniformly labeled, samples representing "leakage" from geranyl pyrophosphate that was destined for biosynthesis of higher terpenoids; the later samples representing the essential oil pool. The essential oil synthesizing site may have contained a large endogenous pool of DMAPP which became labeled slowly. Thus, geraniol synthesized at this site would initially be labeled preferentially in the IPP moiety, but later the DMAPP pool would become labeled. With increasing time, one would expect to find the labeled DMAPP pool reacting with newly synthesized unlabeled IPP, thus reversing the pattern of preferential labeling. The changes observed by Wuu and Baisted between the 12-hour and 24-hour labeling patterns may represent this trend.

With the above discussion in mind it is possible to review certain anomalous results and to provide some advice to those contemplating study of the biosynthesis of lower terpenes. Although preferential labeling has now been demonstrated in several species, equivalent labeling of leaf mono- and sesquiterpenes derived from exogenous MVA-2-^{14}C has been reported in a few instances, e.g., linalool in *Cinnamomum camphora* by Suga *et al.* (1971); citral in *Eucalyptus steigeriana* by Neethling *et al.* (1963); and tutin in *Coriaria japonica* by Corbella *et al.* (1969). However, in these cases, the conditions of the experiments (e.g., drastic isolation techniques and extended incorporation periods) were such that the effects of a DMAPP pool might have easily gone undetected. For example, steam distillation of plant material to remove essential oil may readily convert small quantities of geranyl pyrophosphate to linalool [the major acid hydrolysis product of geranyl pyrophosphate is linalool (Rittersdorf, 1965), and from the biosynthesis of squalene in peppermint it can be inferred that relatively high levels of equivalently labeled geranyl pyrophosphate are generated for triterpene biosynthesis, while relatively little preferentially labeled geranyl pyrophosphate is generated for mono- and sesquiterpene biosynthesis during the same time period]. Linalool isolated by steam distillation from *Cinnamomum camphora* was shown to have incorporated a small amount of exogenous MVA-2-^{14}C label (0.012 percent) and to be equivalently labeled (Suga *et al.*, 1971). The obvious question is, does this result reflect *in vivo* essential oil synthesis, or does it represent geranyl pyrophosphate intended for triterpene biosynthesis? We cannot answer the question, but only suggest that isolation techniques can easily produce artifacts which may lead to erroneous conclusions.

Similarly, long incorporation periods are another possible source of error. During the biosynthesis of terpenes from exogenous precursors a number of simultaneous events are taking place: labeled terpenes are turning over, endogenous pools are being diluted with exogenous label; and synthesis, degradation, and resynthesis of labeled compounds are taking place. This is true even of MVA-^{14}C, which has been shown in several cases to be a more efficient precursor of CO_2 (Battu and Youngken, 1966) and acetyl units (Banthorpe *et al.*, 1970) than of monoterpenes. The longer the incorporation period, the greater will be the combined effects of these processes, and the greater will be the "scrambling" of the original label. This is particularly true when "continuous" rather than "pulse" feeding of the radioactive precursor is employed.

Clearly isotope incorporation data must be evaluated very carefully. Mere incorporation of label from a postulated precursor does not prove a specific biosynthetic sequence, especially when working with trace constituents or low levels of incorporation. Degradations, after relatively short incorporation times, are required to demonstrate specific incorporation. A time course of labeling should be carried out before such studies are started, and the product should, if possible, be degraded at several points of the time-course.

It was stated above that certain monoterpenes formed in flowers were found to be equivalently labeled. In these cases, equivalent labeling does not appear to be attributable to the conditions of the experiment, but rather to the probability that the site of synthesis is fundamentally different physiologically than in the other tissues studied. For example, rose petals incorporate up to 22 percent of applied *R*-MVA-2-^{14}C into equivalently labeled geraniol and nerol within 1 hour (Francis and O'Connell, 1969; Francis *et al.*, 1970), and chrysanthemum ovules incorporate over 2 percent of applied *R*-MVA-2-^{14}C (in 24 hours) into chrysanthemum acids that are equivalently labeled in both C_5 units (Crowley *et al.*, 1962; Godin *et al.*,

1963). The example of the chrysanthemum acids may not be relevant, however, as these irregular monoterpenes may be formed by an unusual condensation of two DMAPP units rather than by the more classical condensation of DMAPP with IPP (Godin *et al.*, 1963). The example of rose petals is certainly relevant, and it appears that a DMAPP pool does not participate in the biosynthesis of monoterpenes in this tissue. It is also possible that a DMAPP pool exists but is rapidly diluted with the relatively large amount of MVA-2-^{14}C label incorporated. In either case, rose petals appear to be quite unlike other tissues studied (e.g., peppermint leaves).

Compartmentalization and Energy Status of Biosynthetic Sites

General Considerations

A number of explanations have been put forward to account for the low incorporation of exogenous MVA into mono- and sesquiterpenes, including competition from other terpenoid pathways for labeled precursor, the derivation of the DMAPP portion of the terpene molecule from an internal carbon pool, and compartmentalization of mono- and sesquiterpene biosynthesis in sites not readily accessible to exogenous MVA. Endogenous DMAPP pools and competition for substrate may operate as expressions of such compartmentalization. Certainly, compartmentalization of biosynthetic mechanisms and their related intermediates exists in plants in a high order of complexity (Oaks and Bidwell, 1970) and, in fact, a compartmentalization phenomenon similar to that suggested for the biosynthesis of lower terpenes is well established in the biosynthesis of higher terpenoids in green leaves (Rogers *et al.*, 1968). Thus, the site of sterol and triterpene biosynthesis in leaves is extrachloroplastidic and is relatively accessible to exogenous MVA-2-^{14}C, while the biosynthesis of carotenoids occurs only within chloroplasts, preferentially utilizing endogenous MVA derived from photosynthetically fixed CO_2. The main aspects of this compartmentalization appear to be the segregation of enzymes and the relative impermeability of intracellular membranes (e.g., chloroplast membranes) to terpenes and their immediate precursors, including MVA. We suggested (Loomis, 1967) that a similar compartmentalization, within the oil glands, operates in the biosynthesis of monoterpenes, and it now appears that sesquiterpene biosynthesis in peppermint is also compartmentalized. Recent studies using sensitive gas chromatographic techniques have now demonstrated that oil glands from several species, including peppermint, actually contain the mono- and sesquiterpenes characteristic of the plant (Amelunxen and Arbeiter, 1967; Hefendehl, 1967, 1968; Sticher and Flück,

1968; Amelunxen *et al.*, 1969; Malingre *et al.*, 1969; Henderson *et al.*, 1970). Several enzymes, including mevalonic kinase, have been detected in isolated glands of hops (*Humulus lupulus*) (Shine and Loomis, unpublished observations). Thus the oil glands appear to be a likely site of synthesis, and the morphology of most types of oil glands is such as to suggest a degree of isolation from the rest of the plant and from the atmosphere. For example, peppermint has two types of oil-secreting epidermal trichomes, both heavily cutinized, and attached to the leaf by a single stalk cell.

Tracer studies described above suggested that monoterpenes and sesquiterpenes are synthesized at different sites, and that the sesquiterpene site is the less isolated of the two. Glandular scales isolated from peppermint cuttings fed MVA-^{14}C contain much more hexane-soluble ^{14}C-label than might be expected on the basis of MVA incorporation into monoterpenes and sesquiterpenes alone (Knox and Loomis, unpublished observations). Thus, if the glands are the site of biosynthesis of mono- and sesquiterpenes, compartmentalization of terpene metabolism and the presence of multiple biosynthetic sites are likely to exist within the glands themselves.

The ready incorporation of MVA-^{14}C into rose petal monoterpenes suggests that rose petals do not have the same kind of strict compartmentalization of terpene metabolism as other tissues studied. This physiological uniqueness appears to be related to the presence of a unique type of oil gland, a glandular epidermis, in flower petals.

Further discussion of oil gland function must await the review of morphology of secretory structures in the next section. It is sufficient to state at this point that oil glands appear to be of fundamental importance in the biochemistry and physiology of mono- and sesquiterpenes

Energy Status of Biosynthetic Sites

If, as we suggest, the sites of mono- and sesquiterpene biosynthesis are, in most cases, compartmentalized within the oil glands, then a number of factors may be operative in addition to the physical exclusion of MVA from the biosynthetic sites. Burmeister and von Guttenberg (1960) have studied the accumulation of essential oil under low oxygen conditions and under the influence of various metabolic inhibitors, and have suggested that the production of essential oil is an adaptation to limited oxygen supply, analogous to alcoholic fermentation in yeast. *In vivo* biosynthesis of MVA from glucose via acetyl-CoA yields, concomitantly, ATP and reduced pyridine nucleotides (i.e., energy), both of which are required for further biosynthesis and metabolism. However, when exogenous MVA-2-^{14}C is fed alone, necessary cofactors must still be generated endogenously for further terpene biosynthesis. This requirement may pose no problem in most pathways of intermediary metabolism, but for biosynthesis taking place within an isolated organ, such as an oil gland, where photosynthate may not be readily available and where primarily fermentative mechanisms may be operative, the generation of cofactors required for the utilization of exogenous MVA may pose a formidable metabolic obstacle. Such a situation would make the cells of the oil glands extremely sensitive to the types and amounts of fermentable substrate (e.g., sucrose) available to them from adjacent cells. To test this hypothesis, we studied the effects of coadministration of unlabeled substrates on the incorporation of MVA-2-^{14}C into mono- and sesquiterpenes of peppermint (Croteau *et al.*, 1972b).

Figure 10 shows the results of one type of study that was carried out, in which the effects of sucrose on MVA incorporation were tested. Aqueous solutions of MVA-2-^{14}C were fed to individual peppermint cuttings through the cut stems, either alone or with 1 μmole of unlabeled sucrose or mannitol. When MVA-2-^{14}C was fed alone, the time course of terpene labeling was similar to that previously observed (Fig. 3): sesquiterpenes were labeled much more effectively than monoterpenes, while both classes of terpenes

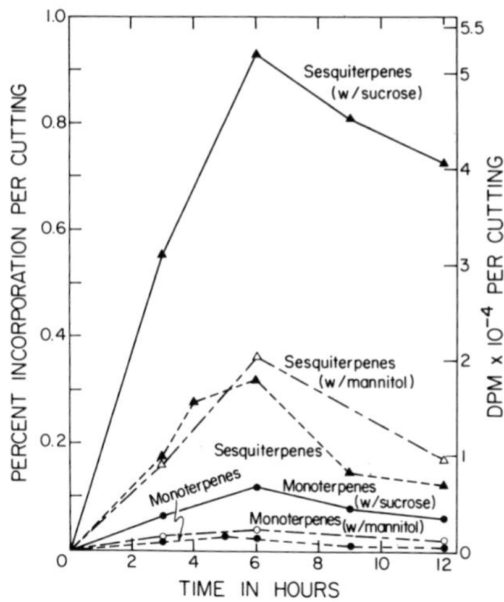

FIG. 10. Time-course of labeling of peppermint mono- and sesquiterpenes from mevalonic acid-2^{14}C with and without added sucrose or mannitol. Data presented are from a single set of matched cuttings fed 1 μmole of *RS*-MVA-2-^{14}C per cutting with or without 1 μmole of added sucrose of mannitol. Data points for mono- and sesquiterpenes represent the sums of activities in gas radiochromatographic peaks. Time is from start of MVA feeding, and percent incorporation is calculated by assuming that only *R*-MVA is physiologically active. From Croteau *et al.* (1972b).

acquired maximum label in about 6 hours and then lost label. Coadministration of 1 μmole of unlabeled mannitol did not appreciably affect the results, whereas 1 μmole of unlabeled sucrose had a definite stimulatory effect on the incorporation of MVA into both mono- and sesquiterpenes at all time periods studied. The fact that mannitol did not stimulate MVA incorporation indicates that the stimulation by sucrose is not due to osmotic effects. Sucrose stimulated the incorporation of MVA into all the individual terpenes that could be analyzed by gas radiochromatography (representing over 90 percent of the essential oil), and so the mono- and sesquiterpenes are plotted as classes (i.e., the sums of gas radiochromatographic peaks) in Fig. 10 rather than as individual components.

The effect of sucrose level (up to 10 μmole) on 6-hour MVA incorporations was then studied. The degree of MVA incorporation into monoterpenes increased linearly with increasing amounts of sucrose up to about 1 μmole per cutting and then leveled off, while the incorporation of MVA into sesquiterpenes reached its maximum at a lower sucrose level, between 0.25 and 0.50 μmole per cutting. Thus, the sesquiterpene system appeared to reach saturation at lower levels of exogenous sucrose than the monoterpene system, consistent with the previous suggestion that the sesquiterpene biosynthetic site is less isolated than the monoterpene biosynthetic site. At 10 μmoles of sucrose or mannitol per cutting, MVA incorporation decreased, due presumably to osmotic effects in the relatively small cuttings employed.

The effect of exogenous sucrose on stimulating the biosynthesis of mono- and sesquiterpenes from exogenous MVA might be explained by several factors, including the following: (1) sucrose may function by increasing the permeability of appropriate membranes to MVA or, alternatively, be necessary for the conversion of MVA to a more transportable form; (2) the presence of sucrose may exert an MVA-sparing effect, by reducing either the degradation of MVA or its utilization in alternate pathways; or (3) the biosynthesis of mono- and sesquiterpenes may occur in an energy-deficient anaerobic environment such that terpene production takes place only when MVA-2-^{14}C and a readily fermentable substrate are present simultaneously. Although all of these factors may operate in producing the sucrose effect, the third alternative appears to be the most important. To test this last alternative further, an experiment was carried out in which the effect of unlabeled MVA on incorporation of label from sucrose-U-^{14}C into mono- and sesquiterpenes was studied, the rationale being that if sucrose is utilized in stimulating MVA-2-^{14}C incorporation then unlabeled MVA might be expected to suppress sucrose-U-^{14}C incorporation. One micromole of MVA did, in fact, suppress the incorporation of label from 1 μmole of sucrose-U-^{14}C into monoterpenes by about 20 percent and into sesquiterpenes by 10 percent, thus suggesting that products of sucrose catabolism (i.e., ATP and reduced coenzymes) are employed in the utilization of exogenous MVA. Furthermore, by combining the data from experiments on sucrose stimulation of MVA-2-^{14}C incorporation with the data on MVA suppression of sucrose-U-^{14}C incorporation, it could be estimated that the ATP yield from sucrose utilized in the biosynthesis of mono- and sesquiterpenes was roughly 10 μmoles of ATP per 1 μmole of sucrose, consistent with a largely fermentative metabolism of sucrose at the biosynthetic sites.

If the effect of sucrose on MVA-2-^{14}C incorporation is that of providing a suitable energy source at the site of biosynthesis, then the administration of CO_2 in the light during stem feeding of MVA-^{14}C might be expected to produce similar results. When peppermint cuttings were exposed to 5 percent CO_2 in air for 1 hour before or during the feeding of MVA-^{14}C, monoterpenes incorporated 2–5 times as much label as in controls run in air. With 5 percent CO_2, sesquiterpenes incorporated about twice as much MVA-^{14}C label as in controls.

The effects of sucrose and CO_2 on the incorporation of MVA into mono- and sesquiterpenes suggests that the *in vivo* biosynthesis of these compounds may be directly influenced by the presence of sucrose or equivalent products of photosynthesis, which might in turn be controlled by the balance between photosynthesis and the utilization of photosynthate. This balance appears to influence monoterpene biosynthesis in long-term experiments with intact peppermint plants and may be related to the apparent diurnal fluctuation of essential oil observed in some species. We have also found that peppermint cuttings taken from plants at the end of the normal light period (when they should contain the highest levels of accumulated photosynthate) incorporate more MVA label into mono- and sesquiterpenes and show slower turnover than cuttings taken from plants at the beginning of the light period. Thus, abundance of photosynthate appears generally to promote terpene synthesis and/or reduce terpene turnover.

The finding that peppermint cuttings require exogenous sucrose for effective utilization of exogenous MVA-2-^{14}C in the biosynthesis of mono- and sesquiterpenes gives additional support to the concept that mono- and sesquiterpenes are produced at sites relatively isolated from the rest of the plant and further suggests that these sites are located in an energy-deficient, fermentative environment, therefore making metabolic activity greatly dependent on the level of fermentable substrate such as sucrose available from adjacent cells. Thus, the low incorporation of label from exogenous MVA-^{14}C into mono- and sesquiterpenes in many plants may result from two separate effects of compartmentalization: the relative inaccessibility of the glandular biosynthetic site to precursor MVA-^{14}C, and the inability to effectively utilize precursor MVA-^{14}C that does reach the site because of an inherent energy deficiency.

Morphology of Secretory Structures

The picture that emerges from tracer investigations of terpenoid biosynthesis in plants is one of multiple compartmentalized sites, each site producing characteristic terpenoid compounds and having its own unique physiology. The tracer evidence is, however, circumstantial and tells us little about the location or physical nature of these biosynthetic sites. If the postulated compartments exist, they must correspond in some way to morphological entities—tissues, cells, or intracellular structures or regions.

Chloroplasts

Compartmentalization between chloroplastidic and extrachloroplastidic terpenoid synthesis in green leaves is well established (Rogers *et al.*, 1968), and we have already suggested above that the hemiterpene isoprene is synthesized in the chloroplasts. What can we now say about the sites of synthesis of mono- and sesquiterpenes?

Oil Glands

Accumulation or secretion of any quantity of mono- or sesquiterpenes is generally, perhaps always, associated with the presence of recognizable glandular structures: notably oil cells, glandular hairs, oil or resin ducts, or glandular epidermis. Attempts to demonstrate essential oil synthesis in plant tissue cultures have not succeeded, and it has been suggested that this was because the cultures did not form oil glands (Carew and Staba, 1965; Becker, 1970). It is commonly supposed that the essential oil components are produced within the "secretory cells" of the oil glands.

The anatomy of plant secretory structures has been studied extensively with the light microscope. Unfortunately, much of this literature is old, or in obscure journals, or both, so that access to it can be difficult, but there are several useful reviews (Weichsel, 1956; Kisser, 1958; Uphof, 1962; Hummel and Staesche, 1962; Esau, 1965). Recently the electron microscope has produced a revival of interest in plant secretory structures (Schnepf, 1969), but primarily in glands that secrete hydrophilic substances rather than in oil glands.

However, several types of oil glands from several plant species have now been examined by electron microscopy. The results generally confirm the earlier observations and provide some ultrastructural detail that was not previously available. Although the number of observations to date is very limited, certain tentative generalizations appear to be warranted. The oil glands that have been examined by high resolution electron microscopy represent diverse morphological types, but in all of them the secretory cells are characterized by very dense protoplasm, at least initially, and by progressive degeneration of cytoplasmic membrane structures during the development of the cells. In the "mature" gland these membranes are either very disorganized or completely lacking. Essential oil production occurs during the time that the membrane structures are degenerating, but it is not yet clear whether it continues after the cells "mature." This developmental picture contrasts sharply with that observed in glands that secrete hydrophilic substances, which are typically characterized by an abundance of mitochondria and dictyosomes in the fully developed cells (Schnepf, 1969).

Within this general pattern of membrane degeneration, the plant oil glands appear to show one or the other of two patterns of cell development. One type [*Mentha* (Amelunxen, 1964, 1965; Amelunxen *et al.*, 1969); *Cleome* (Amelunxen and Arbeiter, 1969)] is characterized, in early stages of cell development, by a highly developed endoplasmic reticulum, and few plastids. The essential oil apparently is formed in the ground plasm and first appears as many small osmium-staining areas or droplets. The glandular epidermis of rose petals may belong to this group (Stubbs and Francis, 1971), although the published observations are not sufficiently detailed to allow of judging with certainty. The other type of gland [*Citrus* (Heinrich, 1966); *Acorus* (Amelunxen and Gronau, 1969); *Dictamnus* (Amelunxen and Arbeiter, 1967)] is characterized by an abundance of plastids in the secretory cells, and the plastids appear to be the site of essential oil and resin synthesis. The internal glandular hairs of patchouli (*Pogostemon cablin*) apparently belong to this group as well (Henderson *et al.*, 1970), although no detailed developmental studies have yet been reported. The secretory cells of *Pinus* resin ducts may represent a combination of the two types. The secretion appears to originate in degenerate plastids which are sheathed by endoplasmic reticulum (Wooding and Northcote, 1965).

Amelunxen and Arbeiter (1967) and Amelunxen (1971) have suggested that the enzymes that produce mono- and sesquiterpenes in secretory cells arise alternatively from the systems in the ground plasm and endoplasmic reticulum that synthesize, for example, sterols, or from the systems that synthesize carotenes and other terpenoids in the plastids. Membrane structures from the endoplasmic reticulum or the plastids, respectively, or perhaps both simultaneously, would provide the compartmentalization necessary to protect the living protoplasm from these secretions. From the limited observations available it appears that the "plastid" type of secretory cell may be characteristic of glands that originate from mesophyll cells, and the "endoplasmic reticulum" type characteristic of glands that originate from epidermal cells.

From this point of view more detailed studies of glandular epidermises and of the secretory hairs of patchouli will be of great interest. Patchouli has both internal (mesophyll) and external (epidermal) secretory hairs, which are very similar in gross morphology (Solereder, 1907; Henderson *et al.*, 1970). The internal hairs may be of the "plastid" type (based on limited observations), but the epidermal hairs have not yet been examined by electron microscopy.

Patchouli is not unique in having more than one type of secretory structure. *Cnicus benedictus* L., for example, has both oil-secreting epidermal hairs and internal oil ducts (Rosenthaler and Stadler, 1908). *Dictamnus albus* has external glandular hairs and "spouting glands" in addition to internal oil glands (Amelunxen and Arbeiter, 1967). Peppermint has two types of external glandular hairs or trichomes: 3-celled "glandular hairs" with one secretory cell (Amelunxen, 1964) and 10-celled "glandular scales" with 8 secretory cells (Amelunxen, 1965). Both have been examined in considerable detail by electron microscopy (Amelunxen, 1964, 1965; Amelunxen *et al.*, 1969). The two types of peppermint glands are quite different from each other cytologically as well as morphologically. Peppermint glandular scales are unique among the oil glands examined in the extent of degeneration of internal membrane structures at a very early stage of leaf development (Amelunxen, 1965; Loomis, 1967). Peppermint glandular hairs are also unusual compared to other oil glands in having numerous well developed dictyosomes (Amelunxen, 1964; Amelunxen *et al.*, 1969), although here also there is a general disintegration of membrane structures in the secretory cell. Both kinds of peppermint glands are of epidermal origin, and both are of the endoplasmic reticulum type.

Grahle (1955) reported estimates of 1700 to 11,900 oil glands per leaf in peppermint, varying with the size of the leaf. Leaves from the sixth to the tenth node all had over 10,000 glands per leaf. Lemli (1963) reported as many as 17,500 glands per leaf at the tenth node. These values apparently refer only to the glandular scales. There was no mention of the 3-celled glandular hairs before Amelunxen's reports, and apparently their glandular nature was not recognized. However, these 3-celled hairs show clearly in Lemli's photographs and appear to be at least as numerous as the 10-celled scales. Thus, by conservative estimate each peppermint leaf bears from 3000 to 30,000 glandular trichomes on its epidermis. That each of these glands is a physiological unit is indicated by histochemical tests performed by Lemli, using a color test for menthofuran. In a photographic field that shows 6 glandular scales, the staining ranged from very dark to very light, and no two scales stained to the same degree. Lemli also observed that the glandular scales did not all fill with oil simultaneously and suggested that those with high menthofuran content had filled earlier.

Peppermint glandular scales and glandular hairs differ fundamentally from each other cytologically, as cited above. They apparently differ physiologically as well. Gas chromatographic analyses of glands isolated from young leaves (<1.5 cm long) (Amelunxen *et al.*, 1969) indicated that the 10-celled scales contained a very "mature" oil in which menthol and menthyl acetate predominated, while the 3-celled hairs contained an "immature" oil, with high menthone. This correlates with the fact that the scales apparently "mature" and lose their internal structure earlier than the hairs. The essential oils are strongly osmiophilic and hence can be seen with the electron microscope (Amelunxen and Arbeiter, 1967; Amelunxen *et al.*, 1969). The oil of the 3-celled hairs of peppermint is contained in many small vacuoles in the cytoplasm. The essential oil of the glandular scales is at first localized in cytoplasmic vacuoles and then is secreted into a subcuticular space.

Extraglandular Secretion

There is direct evidence that essential oils accumulate in the glandular secretory cells. It seems likely that they are also synthesized there, but there are several indications that the oil glands may not be the only sites of essential oil synthesis and accumulation.

Early light microscope observations by several plant anatomists (cited in Uphof, 1962, p. 140) indicated that in some species secretory activity extends from the glands into the neighboring epidermal cells, producing "glandular patches." Recent electron microscopic observations also suggest that essential oil synthesis in pine and peppermint may not be confined to the oil glands. Wooding and Northcote (1965) observed "gray osmiophilic vesicles" in both glandular and nonglandular cells of 1-year-old *Pinus pinea* stems and suggested that they represent stages in the synthesis and secretion of essential oil or resin. Recently Amelunxen (1967), observed similar osmium-staining "filament bundles" in young peppermint leaf cells: both in mesophyll, epidermis, and glandular hairs, but not in the glandular scales. He suggested also that these structures represent essential oil

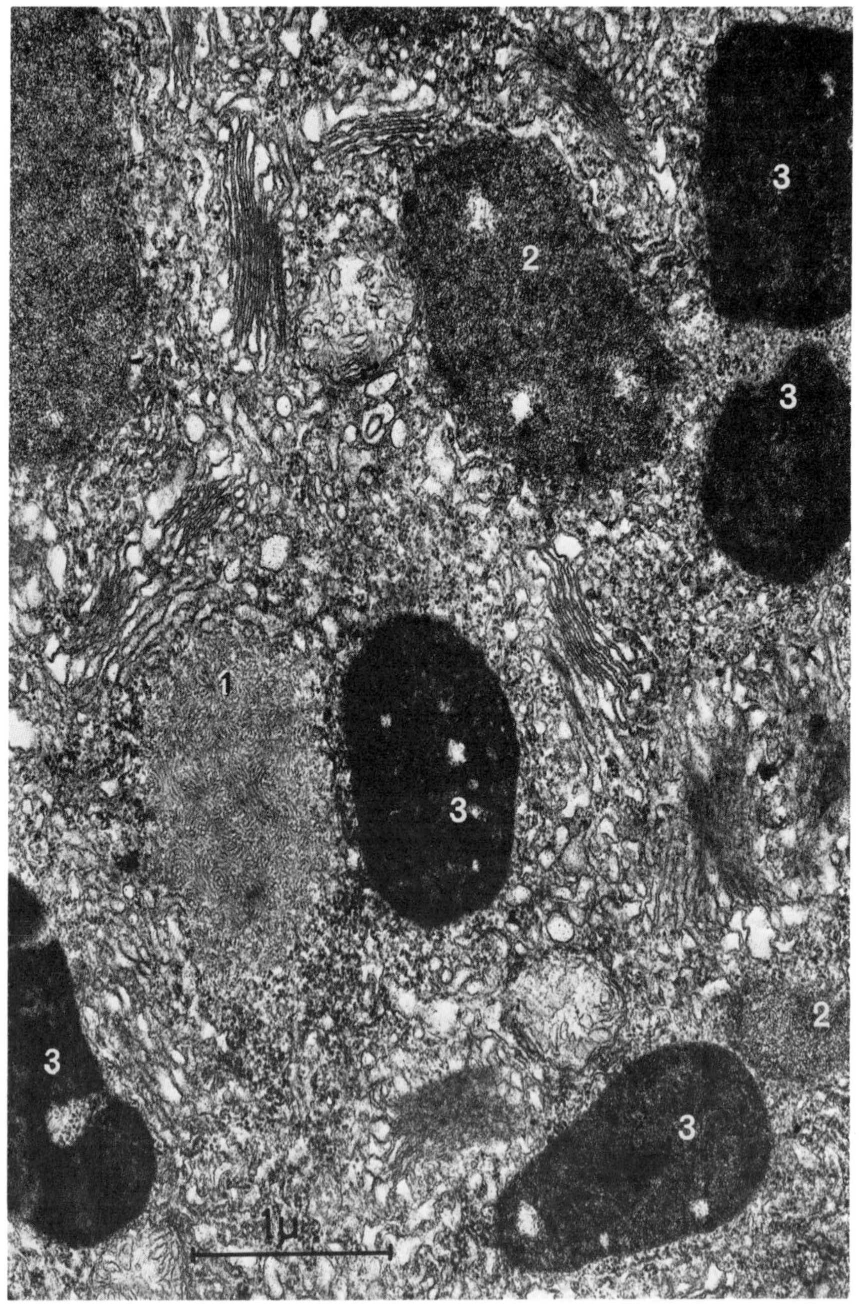

Fig. 11. Unique osmiophilic structures in a peppermint glandular hair Postulated developmental sequence from a "filament bundle area" into an essential oil vacuole is: 1 → 2 → 3. By courtesy of F. Amelunxen (from Amelunxen, 1967).

precursors. Figure 11 shows these structures in a glandular hair, and indicates the postulated sequence of development into an essential oil vacuole.

A further indication of possible essential oil synthesis in sites other than the oil glands comes from comparison of $^{14}CO_2$ incorporation studies and growth chamber studies with Amelunxen's (1964, 1965) developmental studies of peppermint glands. The observed degeneration of structure in the oil gland cells, and the filling of the glands with oil, occur while the leaves are still very young and have in fact hardly started to expand. However, $^{14}CO_2$ studies with peppermint indicated that monoterpene synthesis continued as long as the leaves were still expanding (Battaile and Loomis, 1961). Growth chamber analyses (Burbott and Loomis, 1969) suggested that even fully expanded leaves continued to accumulate essential oil under favorable conditions (cool nights) up until the time of floral initiation. One must conclude that either the oil gland cells function longer than they appear to, or other cells of the leaf contribute to essential oil synthesis and accumulation.

Energy Status of Secretory Sites

Earlier (Loomis, 1967) we suggested that the poor conversion of MVA-^{14}C to monoterpenes in many plant tissues was due to compartmentalization of monoterpene synthesis in sites not accessible to exogenous MVA. Sesquiterpenes would now be included as well. This conclusion was based on findings with species, such as mints and pines, which have glandular hairs or ducts. The ready incorporation of MVA-2-^{14}C into rose petal monoterpenes that was observed subsequently (Francis and O'Connell, 1969) suggested that the rose petal secretory cells were less isolated than other oil-secreting cells. Recent findings, described above, suggest that the sites of mono- and sesquiterpene synthesis in peppermint are not only relatively inaccessible to exogenous MVA but are also energy deficient, due to a limited supply of carbohydrate. Presumably these sites depend largely on anaerobic glycolysis for energy.

Flower petals are unusual not only in the extent to which they utilize MVA-^{14}C for monoterpene synthesis, but also in the anatomy of the glandular cells. The secretory cells of flower petals consist of a glandular epidermis, apparently one of the simplest types of epidermal gland. The epidermal cells expand outward producing papillae, and thus increasing their external surface area (Weichsel, 1956; Mazurkiewicz, 1913; Esau, 1965; Stubbs and Francis, 1971). In some species layers of storage cells have been observed beneath the glandular epidermis, and starch from these cells is depleted during formation of the volatile oil (Esau, 1965). In contrast to glandular hairs, the glandular epidermis seems to be designed for maximum exposure of the secretory cells both to the atmosphere and to the energy-rich mesophyll. It seems unlikely that these cells suffer from deficiency of either oxygen or carbohydrate. From tracer studies it is clear that the monoterpene-synthesizing cells of rose petals take up MVA readily and have abundant energy available for its utilization.

Burmeister and von Guttenberg (1960), referred to essential oil synthesis as a partially anaerobic process and suggested that it is an adaptation to low oxygen supply in the oil glands. The essential factor, however, is that more acetyl-CoA must be formed than can be (or is) oxidized aerobically. This might come about from lack of oxygen, from a deficiency of active mitochondria in the cell, or even from an abundance of carbohydrate in cells with active glycolytic enzymes. In a typical oil-secreting glandular hair there is very likely a deficiency of both oxygen and functional mitochondria. In flower petals, the requisite net surplus of acetyl-CoA might be due to a massive flow of sugars from adjacent cells, such as to overwhelm the mitochondrial respiratory system. Biochemically, what is required to produce mono- and sesquiterpenes is a net surplus of acetyl-CoA production (i.e., more glycolysis than respiration), plus appropriate enzymes to convert acetyl-CoA into terpenoids, and (very probably) membrane structures for intracellular compartmentalization. It appears that there are many different ways of obtaining these conditions, as illustrated by the great diversity of oil gland types.

Plant oil glands have been classified for many years on the basis of morphology, as glandular hairs, ducts, etc. We have cited Amelunxen's suggestion that they are of two cytological types: "plastid glands" and "endoplasmic reticulum glands." It now appears that there may be two physiological types of oil glands as well: (1) isolated, energy-poor glands, probably the most common (as in mint), and (2) energy-rich glands, probably in good communication with adjacent cells and the atmosphere (as in rose petals).* Finally, in a plant with the genetic capacity to produce essential oil in oil glands, it is not inconceivable that this capacity could be expressed on occasion in other cells of the plant.

Localization of Biosynthetic Compartments

At this point we would have liked to correlate further the various postulated biochemical compartments with morphological entities. Unfortu-

* For further discussion of the nature of the secretion of floral scents, both fragrant and foul, see Vogel (1962). Vogel referred to the unique floral scent glands as "osmophors" and characterized them as having abundant supplies of carbohydrate or lipid and good access of air to the glandular cells—much as suggested here. Unfortunately, at the time this text was written, we had seen only quotations from Vogel's paper (Esau, 1965).

nately, the present state of knowledge in this area is only sufficient to allow a few final speculations. Both growth chamber studies and tracer studies with ^{14}C-labeled precursors have suggested that mono- and sesquiterpenes exist in the plant partially in relatively stable storage pools, and partially in metabolic pools capable of very rapid turnover. In morphological terms, one would expect essential oil in the secretory spaces of glandular structures to be relatively inert metabolically. Essential oil within the glandular cells is probably very active metabolically as long as it is not in vacuoles, and perhaps moderately active after it is secreted into vacuoles. Any essential oil that might occur in cells other than the gland cells would probably be highly metabolically active and would accumulate only under conditions where there is a luxuriant supply of photosynthate. For example, tracer evidence has shown that the newly formed terpenes are turned over very rapidly.

Other tracer studies have indicated that mono- and sesquiterpenes are produced at separate biosynthetic sites differing in their degrees of isolation, and have suggested that, in many cases, internal DMAPP pools participate in the biosynthesis of both mono- and sesquiterpenes. Electron micrography of oil glands and of other cells of several essential oil-producing species shows a structural complexity that could easily provide sufficient "theoretical compartments" to account for many endogenous pools and many biosynthetic sites with different degrees of accessibility.

Conclusions

Recent studies have revealed some very interesting aspects of the biochemistry of lower terpenoids and have called into question a number of preconceived notions about these compounds. Thus, the existence of isoprene as an important plant product has now been demonstrated. We have suggested that isoprene is synthesized in chloroplasts, from products of photosynthetic CO_2 fixation. Mono- and sesquiterpenes have been shown to be metabolically active and are not, as was once supposed, inert metabolic end products. Monoterpenes and sesquiterpenes appear, at least in peppermint, to be biosynthesized at different sites.

Many features of mono- and sesquiterpene biosynthesis appear to be intimately related to the compartmentalization of the biosynthetic sites. The concept of compartmentalization thus assumes central importance in mono- and sesquiterpene biosynthesis. The exact nature of this compartmentalization has not yet been elucidated, although it appears to be associated with the presence of specialized secretory structures. Studies on the morphology of secretory structures have revealed considerable ultrastructural complexity and indicate the existence of a number of possible biosynthetic compartments. It has been suggested that compartmentalization of biosynthesis and metabolism occurs in the oil glands, and perhaps in other cells, and within these cells in many cytoplasmic regions separated by membranes that arise from plastids or from endoplasmic reticulum. It also appears that there may be two fundamental physiological types of oil gland secretory sites: isolated energy-deficient sites (as in peppermint), and nonisolated energy-rich sites (as in rose petals). The former utilize exogenous labeled precursors poorly and incorporate label preferentially into IPP-derived moieties, suggesting the presence of an endogenous DMAPP pool. The latter utilize exogenous precursors readily and appear not to have a DMAPP pool. In peppermint, monoterpene and sesquiterpene sites both appear to be isolated and energy-deficient, but not to the same degree.

It is only through the combination of chemical, biochemical, physiological, and morphological studies that we can hope to gain a basic understanding of the biochemistry and physiology of lower terpenoids, and to integrate the lower terpenoids into the overall metabolism of the plant. With the advent of newer techniques such as GC-MS and coupled GC-radiocounting, and improved techniques in electron microscopy and plant enzymology, the biochemistry and physiology of lower terpenoids can now, finally, be studied adequately, and we can look forward to exicting progress in this field.

Note Added in Proof

For additional studies of the ultrastructure of plant oil glands, see Schnepf (1972) and references cited therein.

Acknowledgments

We are grateful to the National Science Foundation for support in the form of a research grant (GB-25593) and to the National Institute of General Medical Sciences, U.S. Public Health Service, for a postdoctoral fellowship (GM 47070) to R.C.

References

Agranoff, B. W., H. Eggerer, U. Henning, and F. Lynen. 1960. *J. Biol. Chem.* **235**:326
Amelunxen, F. 1964. *Planta Med.* **12**:121.
Amelunxen, F. 1965. *Planta Med.* **13**:457.
Amelunxen, F. 1967. *Planta Med.* **15**:32.
Amelunxen, F. 1971. Personal communication.
Amelunxen, F., and H. Arbeiter. 1967. *Z. Pflanzenphysiol.* **58**:49.

Amelunxen, F., and H. Arbeiter. 1969. *Z. Pflanzenphysiol.* **61**:73.
Amelunxen, F., and G. Gronau. 1969. *Z. Pflanzenphysiol.* **60**:156.
Amelunxen, F., T. Wahlig, and H. Arbeiter. 1969. *Z. Pflanzenphysiol.* **61**:68.
Banthorpe, D. V., and D. Baxendale. 1970. *J. Chem. Soc. C* p. 2694.
Banthorpe, D. V., and A. Wirz-Justice. 1969. *J. Chem. Soc. C* p. 541.
Banthorpe, D. V., and Charlwood, B. V. 1971. *Nature New Biol.* (*London*) **231**:285.
Banthorpe, D. V., J. Mann, and K. W. Turnbull. 1970. *J. Chem. Soc. C* p. 2689.
Banthorpe, D. V., B. V. Charlwood, and M. J. O. Francis 1972. *Chem. Rev.* **72**:115.
Battaile, J., and W. D. Loomis. 1961. *Biochim. Biophys. Acta* **51**:545.
Battu, R. G., and H. W. Youngken, Jr., 1966. *Lloydia* **29**:360.
Becker, H. 1970. *Biochem. Physiol. Pflanz.* **161**:425.
Biollaz, M., and D. Arigoni. 1969. *Chem. Commun.* p. 633.
Bonner, J. 1965. *In* "Plant Biochemistry" (J. Bonner and J. E. Varner, eds.), pp. 665–692. Academic Press, New York.
Burbott, A. J., and W. D. Loomis. 1967. *Plant Physiol.* **42**:20.
Burbott, A. J., and W. D. Loomis. 1969. *Plant Physiol.* **44**:173.
Burmeister, J., and H. von Guttenberg. 1960. *Planta Med.* **8**:1.
Carew, D. P., and E. J. Staba. 1965. *Lloydia* **28**:1.
Corbella, A., P. Gariboldi, G. Jommi, and C. Scolastico. 1969. *Chem. Commun.* p. 634.
Croteau, R., and W. D. Loomis. 1972a. *Phytochemistry* **11**:1055.
Croteau, R., and W. D. Loomis. 1972b. *Phytochemistry* (submitted for publication).
Croteau, R., A. J. Burbott, and W. D. Loomis. 1972a. *Phytochemistry.* **11**:2459.
Croteau, R., A. J. Burbott, and W. D. Loomis. 1972b. *Phytochemistry.* **11**:2937.
Crowley, M. P., P. J. Godin, H. S. Inglis, M. Snarey, and E. M. Thain. 1962. *Biochim. Biophys. Acta* **60**:312.
Devon, T. K., and A. I. Scott. 1972. "Handbook of Naturally Occurring Compounds," Vol. II, Terpenes. Academic Press, New York.
Esau, K. 1965. "Plant Anatomy." Wiley, New York.
Francis, M. J. O. 1971. *In* "Aspects of Terpenoid Chemistry and Biochemistry" (T. W. Goodwin, ed.), pp. 29–51. Academic Press, New York.
Francis, M. J. O., and C. Allcock. 1969. *Phytochemistry* **8**:1339.
Francis, M. J. O., and M. O'Connell. 1969. *Phytochemistry* **8**:1705.
Francis, M. J. O., D. V. Banthorpe, and G. N. J. Le Patourel. 1970. *Nature* (*London*) **228**:1005.
Godin, P. J., H. S. Inglis, M. Snarey, and E. M. Thain. 1963. *J. Chem. Soc. London* p. 5878.
Grahle, A. 1955. *Pharmazie* **10**:494.
Green, T. R. 1971. Ph.D. Thesis, Oregon State University, Corvallis, Oregon.
Green, T. R., and D. J. Baisted. 1971. *Biochem. J.* **125**:1145.
Hefendehl, F. W. 1967. *Naturwissenschaften* **54**:142.
Hefendehl, F. W. 1968. *Riechst., Aromen, Koerperpflegem.* **18**:523. (Cited from *Chem. Abstr.* **71**:53443.)
Hefendehl, F. W., E. W. Underhill, and E. von Rudloff. 1967. *Phytochemistry* **6**:823.
Heinrich, G. 1966. *Flora* (*Jena*), *Abt. A* **156**:451.
Henderson, W., J. W. Hart, P. How, and J. Judge. 1970. *Phytochemistry* **9**:1219.
Herout, V. 1971. *In* "Aspects of Terpenoid Chemistry and Biochemistry" (T. W. Goodwin, ed.), pp. 53–94. Academic Press, New York.
Holloway, P. W., and G. Popják. 1967. *Biochem. J.* **104**:57.
Horodysky, A. G., G. R. Waller, and E. J. Eisenbraun. 1969. *J. Biol. Chem.* **244**:3110.

Hummel, K., and K. Staesche. 1962. *In* "Handbuch der Pflanzenanatomie" (W. Zimmermann and P. G. Ozenda, eds.), Vol. IV, Part 5, pp. 209–251. Borntraeger, Berlin.

Jackson, W. A., and R. J. Volk. 1970. *Annu. Rev. Plant Physiol.* **21**:385.

Kisser, J. 1958. *In* "Handbuch der Pflanzenphysiologie" (W. Ruhland, ed.), Vol. X, pp. 91–131. Springer-Verlag, Berlin and New York.

Lemli, J. 1963. *Arch. Med. Angers* **64**:225.

Loomis, W. D. 1967. *In* "Terpenoids in Plants" (J. B. Pridham, ed.), pp. 59–82. Academic Press, New York.

Loomis, W. D., and R. Croteau. *J. Agr. Food Chem.* (submitted for publication).

Malingre, T. M., D. Smith, and S. Batterman. 1969. *Pharm. Weekbl.* **104**:429. (Cited from *Chem. Abstr.* **71**:53444.)

Mazurkiewicz, W. 1913. *Z. Allg. Oesterr. Apoth.-Ver.* **51**:241.

Neethling, L. P., H. G. Reiber, and C. O. Chichester. 1963. *Nat. Conf. Nucl. Energy, Appl. Isotop. Radiat. Proc., Pretoria* pp. 451–54.

Nicholas, H. J. 1962. *J. Biol. Chem.* **237**:1485.

Oaks, A., and R. G. S. Bidwell. 1970. *Annu. Rev. Plant Physiol.* **21**:43.

Parker, W., J. S. Roberts, and R. Ramage. 1967. *Quart. Rev. Chem. Soc.* **21**:331.

Rasmussen, R. A. 1970. *Environ. Sci. Technol.* **4**:667.

Rasmussen, R. A., and C. A. Jones. 1972. *Phytochemistry*, in press. (Paper No. 1469).

Regnier, F. E., G. R. Waller, E. J. Eisenbraun, and H. Auda. 1968. *Phytochemistry* **7**:221.

Rittersdorf, W. 1965. *Angew. Chem., Int. Ed. Engl.* **4**:444.

Robinson, T. 1967. "The Organic Constituents of Higher Plants." Burgess, Minneapolis, Minnesota.

Rogers, L. J., S. P. J. Shah, and T. W. Goodwin. 1968. *Photosynthetica* **2**:184.

Rosenthaler, L., and P. Stadler. 1908. *Arch. Pharm.* (*Weinheim*) **246**:436.

Ruzicka, L. 1959. *Proc. Chem. Soc. London* p. 341.

Sanadze, G. A. 1964. *Sov. Plant Physiol.* **11**:42. (*Fiziol. Rast.* **11**:49.)

Sanadze, G. A. 1966. *Sov. Plant Physiol.* **13**:669. (*Fiziol. Rast.* **13**:753.)

Sanadze, G. A., and G. M. Dolidze. 1961. *Soobshch. Akad. Nauk Gruz. SSR* **27**:747. Cited in Rasmussen. 1970.

Sanadze, G. A., and A. L. Kursanov. 1966. *Sov. Plant Physiol.* **13**:184. (*Fiziol. Rast.* **13**:201).

Sanadze, G. A., G. I. Dzhaiani, and I. M. Tevzadze. 1972. *Fiziol. Rast.* **19**:24. [Cited from Chem. Abstr. **76**:124276 (1972).]

Schnepf, E. 1969. *In* "Sekretion und Exkretion bei Pflanzen" (M. Alfert, H. Bauer, C. V. Harding, W. Sandritter, and P. Sitte, eds.), Protoplasmatologia, Vol. VIII/8, Springer-Verlag, Vienna and New York.

Schnepf, E. 1972. *Biochem. Physiol. Pflanz.* **163**:113.

Schröder, U. 1969. *Pharmazie* **24**:179, 421.

Scora, R. W., and J. D. Mann. 1967. *Lloydia* **30**:236.

Shah, D. H., W. W. Cleland, and J. W. Porter. 1965. *J. Biol. Chem.* **240**:1946.

Solereder, H. 1907. *Arch. Pharm.* (*Weinheim*) **245**:406.

Sticher, O., and H. Flück. 1968. *Pharm. Acta Helv.* **43**:411.

Stubbs, J. M., and M. J. O. Francis. 1971. *Planta Med.* **20**:211.

Suga, T., T. Shishibori, and M. Bukeo. 1971. *Phytochemistry* **10**:2725.

Uphof, J. C. T. 1962. *In* "Handbuch der Pflanzenanatomie" (W. Zimmermann and P. G. Ozenda, eds.), Vol. IV, Part 5, pp. 1–205. Borntraeger, Berlin.

Vogel, S. 1962. *Akad. Wiss. Lit., Mainz, Abh. Math.-Naturwiss. Kl.* No. 10, pp. 599–763 (pages also numbered separately 1–165).
Waller, G. R. 1969. *Progr. Chem. Fats Lipids* **10**:151.
Weichsel, G. 1956. *In* "Die Ätherischen Öle" (E. Gildemeister, F. Hoffmann, and W. Treibs, eds.), Vol I, pp 233–254. Akademie-Verlag, Berlin.
Weiss, B., and H. Flück. 1970. *Pharm. Acta Helv.* **45**:169.
Wooding, F. B. P., and D. H. Northcote. 1965. *J. Ultrastruct. Res.* **13**:233.
Wuu, T. 1971. M.S. Thesis, Oregon State University, Corvallis, Oregon.
Wuu, T., and D. J. Baisted. 1972. *Phytochemistry* (submitted for publication).
Zelitch, I. 1965. *J. Biol. Chem.* **240**:1869.
Zelitch, I. 1969. *Annu. Rev. Plant Physiol.* **20**:329.

GENETIC AND BIOSYNTHETIC RELATIONSHIPS OF MONOTERPENES

ROBERT S. IRVING* and ROBERT P. ADAMS

Botany Department, University of Montana, Missoula, Montana, and Department of Botany and Plant Pathology, Colorado State University, Fort Collins, Colorado

Introduction

In the last one and one-half decades, there has been an increasing interest in secondary plant compounds. Much of this interest has been generated by

* Present address: Biology Department, Louisiana State University at New Orleans, New Orleans, Louisiana.

their applicability to the field of systematic biology, where these compounds have served not only to indicate intrapopulational gene flow and population dynamics, but also to establish phenetic and phylogenetic relationships in various plant groups (Critchfield, 1966, 1967; DeWet and Scott, 1965; DeWet, 1967; Emboden and Lewis, 1967; Fujita, 1965a, b, c; Habeck and Weaver, 1969; Scora, 1967a, b, 1970; Stone, 1964; Stone *et al.*, 1965, 1969; Turner, 1967, 1969). With this increasing interest in secondary compounds, especially monoterpenes and sesquiterpenes, details of their modes of inheritance and formation assume major importance. It is of little use to accord phylogenetic or even phenetic significance to chromatographic profiles or to appraise complex introgressing populations until something is known of the genetic and biosynthetic background of these compounds.

With monoterpenes it is becoming increasingly clear that inter- and intrapopulational variation will reflect at least four major variables (Burbott and Loomis, 1967; Loomis, 1969; Adams, 1970; Firmage and Irving, 1971): (1) individual genetic divergence; (2) the broad environmental conditions of the population and the conditions prevailing at the time of sampling; (3) the ontogenetic and phenological stages; and (4) the techniques of extraction and analysis (e.g., *in vitro* rearrangements). Thus, it is important that genetic variation be isolated from other sources of variation before sound systematic conclusions are drawn. There have been few investigations that dealt directly with the inheritance of monoterpenes. Murray (1960a, b) demonstrated that the difference between two *Mentha* species is due to a single gene with two alleles. Forde (1964) using field variation of monoterpenes presented evidence that the major difference between two species of *Pinus* (high and low α-pinene levels) is simply inherited, but that within one species the range of variation suggested a more complex genetic control. Hanover (1966a) working with oleoresin of F_1's in *Pinus monticola* suggests that most of the monoterpenes of this species are under multigenetic control. One compound, 3-carene, proved to be an exception, being seemingly under the control of a single dominant gene (Hanover, 1966b). Zavarin *et al.* (1969) in the F_1, F_2, and backcross progeny of two *Pinus* species (*P. contorta* and *P. banksiana*) suggests that the monoterpene differences are controlled by a limited number of genes with major effects.

Closely coupled to information on monoterpene inheritance is knowledge of the biosynthetic pathways involved. Indeed, it is often difficult to separate genetics from biosynthesis, and both are prerequisite to sound systematic use of terpene data. Currently, a number of biosynthetic models have been proposed based on labeling experiments (Battu and Youngken, 1966; Scora and Mann, 1967; Sandermann, 1962); structural properties and chemical reactivities (Kremers, 1922; Ruzicka, 1953; Reitsema, 1958), and natural variation (Zavarin, 1970; Zavarin and Cobb, 1970; Zavarin *et al.*, 1971). These numerous schemes although unified in their fundamental premises are, in their details, as diverse as the compounds themselves.

The present work deals with the biosynthesis and the genetics of monoterpenes in experimental populations of three taxa within the mint genus *Hedeoma*. Varying levels of qualitative and quantitative intercorrelations within an experimental F_2 population are used to suggest biosynthetic and genetic relatedness. Further, variation of individual compounds and groups of compounds across the F_2 generation is utilized to provide possible genetic models of inheritance.

The Biology of the *Hedeoma drummondii* Complex

The *Hedeoma drummondii* complex (Labiatae) represents a unique experimental system for addressing questions of monoterpenoid inheritance and biosynthesis. The complex consists of three morphologically defined taxa, *H. drummondii*, *H. reverchonii* var. *reverchonii*, and *H. reverchonii* var. *serpyllifolium*, each possessing a unique terpenoid complement and a high degree of intercrossability. Geographically the complex is centered on the arid limestone outcroppings of the Edwards Plateau region of central Texas (Fig. 1). Here, all three taxa form large sympatric populations on the same chromosome level. The distinguishing morphological characters, although many in number, are, in these regions of sympatry, subtle and intergrading. In fact, there appears to be in each taxa an array of recognizable evolutionary lines or microspecies. It is this recognizability and repetition of variation that had led to much of the classical taxonomic difficulty in dealing with this group. Partial explanation of this complexity lies in their predominant inbreeding behavior which funnels genetic variation into a series of uniform lines. Much of the inter- and intrapopulational variation also appears to be the result of past and present gene exchange between several or all of these taxa. Thus, gene flow, in as yet unknown quantities, is believed to usher in the variation which is then played upon by selection and proliferated by inbreeding. The initial problem was to account for the populational patterns of variation and the origins of the complex itself. Out of these basic systematic and field oriented questions arose the genetic and biosynthetic experiments that are to be discussed. Because morphology did not provide adequate differentiation, the monoterpenes of these taxa were examined. Although the terpenoid data provided the requisite taxonomic separation, concomittantly there had to be greater understanding of the biosynthetic and genetic base of these new characters.

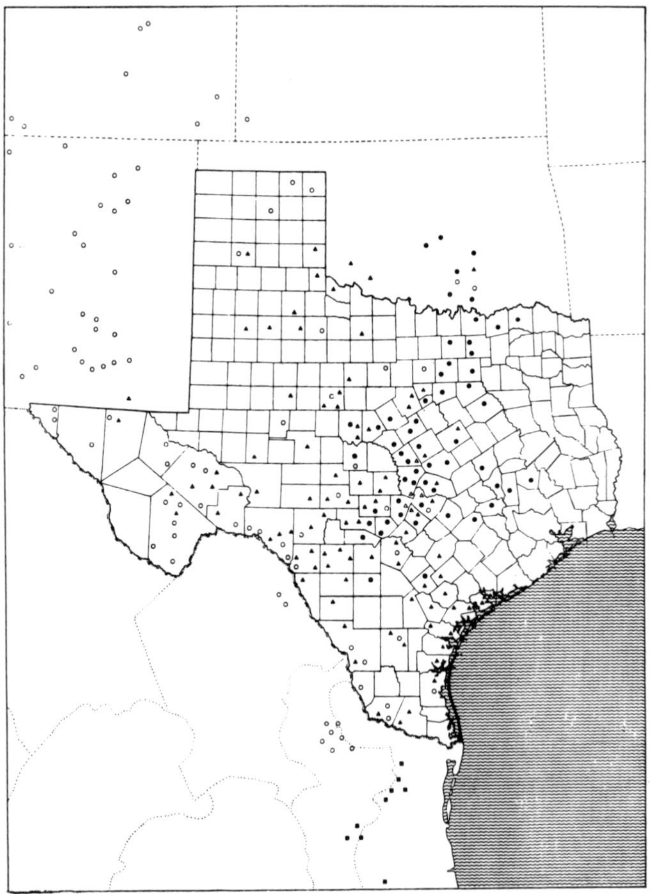

FIG. 1. Distribution of *Hedeome drummondii* complex: *H. drummondii*, open circles; *H. reverchonii* var. *serpyllifolium*, filled triangles; *H. reverchonii* var. *reverchonii*, filled circles.

THE MONOTERPENES OF THE *Hedeoma drummondii* COMPLEX

The salient phytochemical products of the family Labiatae are monoterpenes. Employing standarized, analytical, and preparative gas chromatographic techniques (Irving, 1968), numerous individuals representing several populations were examined as to their monoterpene products (Table 1).

The monoterpenes of *H. drummondii* are ca. 90 percent monocyclic compounds: *d*-limonene, isomenthone, menthone, and pulegone. Although the monocyclic ring theme predominates, *H. drummondii* does not lack the potential to synthesize other ring types in trace amounts (e.g., α-pinene, camphene, sabinene, and myrcene). This pattern, in most of its quantitative

TABLE 1

MONOTERPENES OF THREE PARENTAL TAXA

Monoterpenes	*H. drummondii*	*H. reverchonii* var. *reverchonii*	*H. r.* var *serpyllifolium*
1. α-Pinene	Tr[a]	1.5	8.0
2. Camphene	Tr	1.0	17.0
3. β-Pinene	Tr	Tr	30
4. Sabinene	Tr	Tr	5.0
5. Myrcene	Tr	2.0	2.5
6. *d*-Limonene	1.5	Tr	4.5
7. 1,8-Cineole	Tr	Tr	7.5
8. Terpinene	—	—	2.5
9. Terpinolene	—	Tr	2.0
10. Unknown	—	Tr?	2.0
11. Menthone	1.5	—	—
12. Isomenthone	40.0	—	—
13. Cyclic ketone	1.0	Tr?	Tr
14. Unknown	1.0	Tr?	Tr
15. Tricyclene	—	—	0.5
16. Unknown	1.0	Tr?	4.3
17. Pulegone	50.0	—	—
18. Borneol	—	3.5	26.0
19. Unknown	Tr	1.0	3.0
20. Unknown	—	—	Tr
21. Unknown	?	?	1.0
22. Terpinenol-4	Tr	3.0	1.5
23. Menthol	Tr	—	—
24. *t*-Ocimene	—	4.5	Tr
25. Citronellal	—	22.0	Tr
26. Neral	—	21.0	Tr
27. Geranial	—	31.7	Tr
28. β-Phellandrene	—	Tr	Tr
29. Camphor	—	Tr?	Tr
30. Unknown	—	3.0	—
31. Unknown	Tr	—	—
32. Unknown	—	3.0	—

[a] Tr = < 1 percent.

and all of its qualitative aspects is constant throughout the range of *H. drummondii*. The terpenoid profiles of populations from Montana are remarkably similar to those found in Northern Mexico.

Hedeoma reverchonii var. *serpyllifolium* displays a quite different monoterpene profile with a bicyclic ring theme, the major constituents being borneol, camphene, 1,8-cineole, α- and β-pinenes, tricyclene, and sabinene. As with *H. drummondii*, other ring themes are present in trace amounts.

Although there is quantitative variation found both within and between populations, the overall quantitative and qualitative profile remains invariable.

H. reverchonii var. *reverchonii* displays yet a third distinctive terpenoid pattern with an acyclic monoterpene theme (80–85 percent): neral, geranial, myrcene, *trans*-ocimene, and citronellal. Again there are trace products representing other ring themes, but with the acyclic profile holding from population to population.

Thus, where at the morphological level we have continuous, intergrading characters, at the monoterpene level we have characters that are highly discontinuous and seemingly represent points of major divergence. In light of current biosynthetic views, these three taxa seem to have taken, with the guidance of natural selection, three different biosynthetic routes: acyclic, monocyclic, and bicyclic. Moreover, as we find these patterns intact from taxon to taxon, we might further infer that monoterpenes possess a high degree of heritability and taxonomic reliance. Despite these inferences, however, sound systematic judgments can be made only when nongenetic variation of monoterpenes is reduced and their behavior with gene flow is known. In our work, a series of environmentally controlled breeding experiments was designed to generate the three possible F_1 and F_2 generations.

The Monoterpenes of F_1 Generations

One of the major emphases in micromolecular systematics has been the documentation of hybridization (Alston and Turner, 1963a). The major assumption, stemming primarily from flavanoid data (Alston and Turner, 1963a, b; Smith and Levin, 1963), has been that F_1 hybrids will be recognizable through qualitative and/or quantitative complementation. In monoterpene inheritance this assumption has only been tested quantitatively (Mirov, 1956; Hanover, 1966a; Zavarin *et al.*, 1969).

The experimental *Hedeoma* F_1 hybrids from each of the three combinations of parents yielded the same broad results and confirmed the assumption of complementation. Each F_1 hybrid, although varying in fertility, displayed a complete complementary or additive inheritance pattern of the parental monoterpene products (Table 2). Quantitatively nearly all the monoterpene compounds in the F_1's were intermediate and approached theoretical values based on simple one-to-one addition of the two parental oils. However, in each F_1 type there were transgressive compounds which exceeded the levels found in either of the two parents.

The *H. drummondii* × *H. reverchonni* var *reverchonii* Cross

This cross represents the monocyclic systems of *H. drummondii* in the maternal background of the acyclic *reverchonii* (Table 2). The hybrids were

TABLE 2

MEAN PERCENTAGE OF MONOTERPENES IN F_1 GENERATIONS

Monoterpenes	*Hedeome drummondii* × *H. r.* var. *serpyllifolium*	*H. drummondii* × *H. r.* var. *reverchonii*	*H. r.* var. *serpyllifolium* × *H. r.* var. *reverchonii*
1. α-Pinene	3.4	0.3	3.3
2. Camphene	6.1	0.1	5.2
3. β-Pinene	1.3	0.2	1.0
4. Sabinene	2.6	0.1	3.4
5. Myrcene	1.5	1.1	2.8
6. *d*-Limonene	8.7	5.8	3.0
7. 1,8-Cineole	3.9	Tr	4.1
8. Terpinene	1.2	—	Tr
9. Terpinolene	1.2	—	1.4
10. Unknown	0.4	—	1.7
11. Menthone	11.9	7.8	—
12. Isomenthone	13.5	10.4	—
13. Cyclic ketone	0.3	0.8	Tr
14. Unknown	1.1	0.6	Tr
15. Tricyclene	0.1	—	0.1
16. Unknown	3.8	Tr	5.1
17. Pulegone	16.5	22.5	—
18. Borneol	15.9	1.2	13.6
19. Unknown	3.3	1.5	3.6
20. Unknown	0.3	—	—
21. Unknown	0.8	—	—
22. Terpinenol-4	Tr	3.3	Tr
23. Menthol	Tr	—	—
24. *t*-Ocimene	—	3.4	8.1
25. Citronellal	—	5.3	2.5
26. Neral	—	5.6	10.8
27. Geranial	—	12.5	19.0
28. β-Phellandrene	Tr	0.1	0.1
29. Camphor	Tr	0.5	1.1
30. Unknown	—	11.6	6.4
31. Unknown	—	2.8	—
32. Unknown	—	2.3	2.0

vigorous with an average fertility of 33.2 percent. Their terpenoid profiles were remarkably similar and showed complete complementation; all constituents of both parents were represented in the F_1. Thus, in one plant, and presumably in one gland, there was the coexistence of two disparate pathways: acyclic and monocyclic. Quantitatively, there were three transgressive compounds. Menthone, a trace constituent (0.3 percent) of *H. drum-*

mondii and nonexistent in *H. reverchonii* var. *reverchonii* rose to an average level of 7.8 percent in the F_1 hybrids. Two unidentified trace components (0.5 and 3.5 percent) of *H. reverchonii* var. *reverchonii* rose to 2.9 and 10.6 percent, respectively.

The *H. reverchonii* var. *serpyllifolium* × *H. reverchonii* var *reverchonii* Cross

In this cross a bicyclic system has been incorporated with an acyclic one (Table 2). The hybrids displayed a high degree of fertility (89.0 percent), and as with the previous cross there was qualitative–quantitative complementation and the occurrence of transgressive compounds. However, in this F_1 group there was considerably greater variation which centered principally within the aldehydes, citronellal, neral, and geranial.

The *H. drummondii* × *H. reverchonii* var. *serpyllifolium* Cross

Here a monocyclic system coexists with a bicyclic one with qualitative and quantitative complementation (Table 2). The hybrids themselves were quite healthy despite a lowered fertility of only 16.0 percent. Menthone, entering the system through *H. drummondii,* was noticeably transgressive rising to an average level of 12.8 percent in the F_1 population. Variation between individuals was again centered principally around a carbonyl function in the ketones menthone, isomenthone, and pulegone.

Discussion

With this series of crosses, ostensibly divergent pathways have been shown to coexist within a given plant. At the genetic level, this coexistence is indeed real with every cell of a given F_1 plant possessing the potential for the synthesis of two divergent monoterpene systems. Whether or not this potential is actualized is not as yet known. It may well be that there is a high degree of cellular or intracellular compartmentalization (Loomis, 1972). The transgressive states of several compounds encountered in all the crosses, however, suggest an absence of such compartmentalization, at least at the cellular level. A reasonable model for transgressive character states is that of enzyme complementation (Alston *et al.*, 1965). For example, enzymes for increasing the levels of menthone, or like isomers, may be latent in *H. r. serpyllifolium* and *H. r. reverchonii.* These will then be activated only in the presence of a genome which synthesizes menthone: *H. drummondii.* Such a model would necessitate a decompartmentalized system with enzymatic interchange.

The variation within each F_1 population was primarily with carbonyl functions. Recent biosynthetic models would order such compounds along a reductive sequence (Reitsema, 1958; Loomis, 1969). Such an alignment would remove the products sequentially from their precursor and accord them a greater degree of biosynthetic autonomy. Further, with such a sequence, there would be greater chance of reversibility and sensitivity to environmental change (Loomis, 1969; Firmage, 1971).

The Monoterpenes of F_2 Generations

Monoterpene analysis of experimentally controlled F_2 populations can bring insight into three major areas: (1) the quality and quantity of recombination in an F_1 hybrid of divergent parentage; (2) the biogenetic relationships suggested from the variation; and (3) the genetic background of individual or groups of monoterpene compounds. In our work it was further hoped that these patterns of variation would provide direction for subsequent experimental and field research.

A single F_1 hybrid was chosen from the *H. drummondii* × *H. r. serpyllifolium* cross (monocyclic × bicyclic), and through inbreeding an F_2 population of approximately 300 individuals was established. Of these, 106 were randomly selected for analysis. Of the 45–50 compounds, 21 were singled out for study, the selection based primarily on quantitative "importance" (Table 1). In addition, 10 morphological attributes were examined along with individual plant fertility. The entire generation was subjected to the same environmental regimes as the parents and F_1's and sampling was conducted at comparable developmental stages.

The most difficult appraisal of variation is that of entire patterns, individual to individual. Indeed, it is this type of appraisal which is the heart of the "taxonomic method." However, as the F_2 chromatograms were examined, it was quickly realized that the variation throughout the F_2 generation could be subjectively taxonomized—the patterns were not as variable as one might expect on *a priori* grounds. This conservative variation could be categorized into eight subjective groupings.

The largest group (27.0 percent of the F_2 generation) (Fig. 2) displayed patterns which approximated the qualitative and quantitative patterns of the F_1 hybrid. Morevoer there were two subgroups (8.0 and 19.0 percent, respectively) which also followed the F_1 theme but differed significantly in the levels of pulegone, isomenthone, and menthone. A fourth group was subjectively clustered through depressed levels of the monocyclic ketones and the presence of an unknown ketone at transgressive levels (compound 13). Of particular importance are groups 5 and 6 (22.6 percent) which repre-

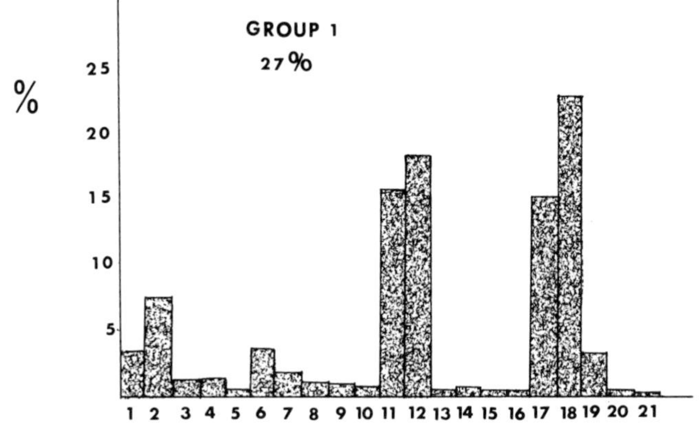
GROUP 1
27%
%
25
20
15
10
5
1 2 3 4 5 6 7 8 9 10 11 12 13 14 15 16 17 18 19 20 21

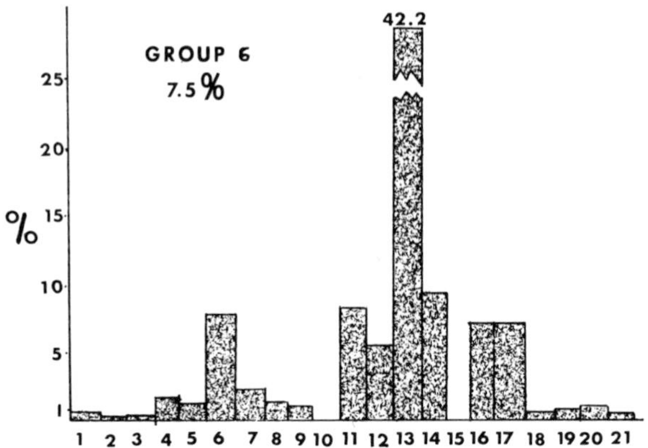
42.2
GROUP 6
7.5%
%
25
20
15
10
5
1
1 2 3 4 5 6 7 8 9 10 11 12 13 14 15 16 17 18 19 20 21

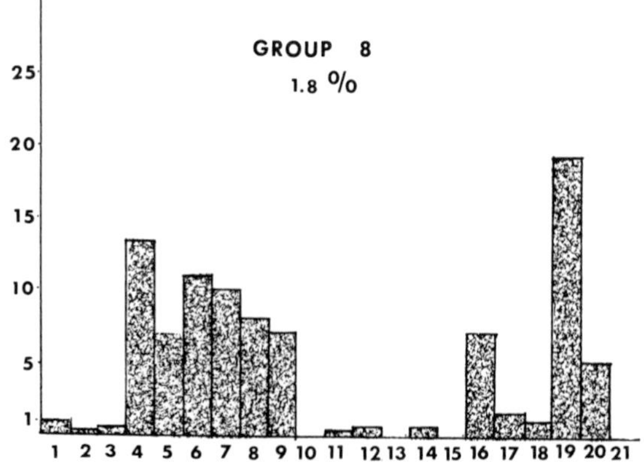
GROUP 8
1.8 %
25
20
15
10
5
1
1 2 3 4 5 6 7 8 9 10 11 12 13 14 15 16 17 18 19 20 21

sent a complete (group 5) or nearly complete (group 6) return to the parental patterns of the *H. drummondii* parent. Group 6 deviated from the *H. drummondii* profile with the presence of the same transgressive ketone present in group 4 (Fig. 3). The monoterpene return to the *H. r.* var. *serpyllifolium* parent (group 7) represented 9.4 percent of the F_2 population. This segregation, back to the parental profiles, was complete both qualitatively and quantitatively. The remaining group, although a small portion of the sample, was instructive in that it represented "radical" recombinants (Fig. 4).

Although these groups are in part subjective, they suggest a high degree of coherency in the monoterpene patterns. Groups of monoterpenes will not breakdown readily through genetic and chromosomal recombination.

To treat the F_2 generation as a variable population and examine individual compounds or groups of compounds out of the individual patterns of variation, two basic techniques were used: (1) intercorrelation analysis which provided correlation coefficients for the 21 monoterpene products, 10 morphological characters, and fertility (Table 3); (2) factor analysis to cluster those characters together which were covarying (Fig. 5). Intercorrelation analysis, as a means of interpreting terpenoid variation in genetic or biosynthetic terms, has been used by several workers (Hanover, 1966a, b; Zavarin, 1970). In an experimental F_2 generation, the correlation coefficients of monoterpenes can assume a high degree of significance as both gene flow and environment have been highly controlled and selective pressures reduced to a minimum. With the above controls, correlations between two monoterpenes, representing concomitantly genetic and biosynthetic relationships, can be one of three intergrading types*: (1) a nonsignificant correlation representing a high degree of genetic and biosynthetic independence; (2) negative correlations representing the synthesis of one compound at the expense of the other or its precursor and, thus, some degree of biogenetic interdepencence; and (3) a large positive correlation signifying a codependence and perhaps a higher degree of common genetic and biosynthetic control.

* For the problem of spurious high correlations through quantification by "percent total oil," see discussion by Zavarin (1970), which we accept.

FIG. 2 (top). Group 1 F_2 pattern, which approximates F_1.

FIG. 3 (middle). Group 6 F_2 pattern, which is essentially a return to *Hedeoma drummondii* except for compound 13 (ketone).

FIG. 4 (bottom). Group 8 F_2 pattern with a high degree of recombination.

TABLE 3

CORRELATION

	1	2	3	4	5	6	7	8	9	10	11	12
1. α-Pinene	1.0	0.94	0.995	0.35	0.32	−0.08	0.36	0.14	0.34	0.66	−0.31	−0.30
2. Camphene		1.0	0.94	0.28	0.24	−0.13	0.27	0.04	0.28	0.73	−0.28	−0.31
3. β-Pinene			1.0	0.30	0.27	−0.11	0.32	0.11	0.29	0.67	−0.29	−0.27
4. Sabinene				1.0	0.99	0.55	0.93	0.75	0.99	0.12	−0.36	−0.47
5. Myrcene					1.0	0.58	0.91	0.76	0.98	0.05	−0.37	−0.48
6. *d*-Limonene						1.0	0.56	0.51	0.54	−0.20	−0.24	−0.39
7. 1,8-Cineole							1.0	0.68	0.93	0.17	−0.34	−0.45
8. Terpinene								1.0	0.75	0.03	−0.24	−0.35
9. Terpinolene									1.0	0.12	−0.36	−0.47
10. Unknown										1.0	−0.06	−0.14
11. Menthone											1.0	0.42
12. Isomenthone												1.0
13. Ketone												
14. Unknown												
15. Tricyclene												
16. Unkown												
17. Pulegone												
18. Borneol												
19. Unknown												
20. Unknown												
21. Unknown												
23. Fertility												
27. Calyx length												

Because of the paucity of information on monoterpene interconversions and the difficulty in separating biosynthesis from genetics, definitive models of biosynthesis cannot be assigned to any of these intercorrelation categories. Moreover, with all schemes of monoterpene interconversion, it is as yet difficult to distinguish the four possible modes of biosynthesis and gene action: (1) nonenzymatic conversions in which the quantitative relationships are the product of reaction rates; (2) nonspecific enzymatic conversions where ubiquitous enzymes will act on a variety of substrates to produce a variety of monoterpene products; (3) specific enzymatic conversions with the enzyme geared to a specific monoterpene substrate and product; and (4) biologically, and perhaps more realistically, *a priori*, a combination of all three. Yet despite these inherent difficulties and the complexity of the system, it may be fruitful to attempt such modeling, fully realizing the risks of oversimplification.

With relatively high negative correlation coefficients two related biosynthetic models are plausible: (1) a sequence or series of interdependent biosynthetic steps; (2) differential enzymatic competitiveness for a common precursor. With a high positive correlation coefficient, a biosynthetic and genetic model based on a common intermediate becomes attractive.

Individual Variation

A synopsis of the individual monoterpene variation (Fig. 6) in the F_2 generation revealed a wide range for each monoterpene, and in many in-

Coefficient Matrix

13	14	15	16	17	18	19	20	21	23	27	
−0.26	−0.12	0.91	−0.04	−0.45	0.84	0.11	−0.19	0.06	−0.03	−0.08	1.
−0.27	−0.15	0.98	−0.07	−0.43	0.93	0.09	−0.22	0.04	−0.06	−0.04	2.
−0.24	−0.09	0.91	−0.09	−0.45	0.83	0.06	−0.23	0.06	−0.04	−0.07	3.
−0.17	−0.18	0.27	0.66	−0.46	0.20	0.85	0.62	0.02	−0.03	−0.06	4.
−0.11	−0.14	0.22	0.68	−0.45	0.13	0.85	0.63	−0.01	−0.01	−0.06	5.
0.18	−0.06	−0.10	0.53	−0.27	−0.16	0.55	0.47	0.13	−0.34	−0.16	6.
−0.18	−0.17	0.25	0.54	−0.45	0.22	0.73	0.48	0.04	−0.12	−0.07	7.
−0.06	−0.15	0.04	0.43	−0.33	0.04	0.62	0.67	−0.00	−0.06	−0.14	8.
−0.17	−0.20	0.26	0.65	−0.45	0.21	0.86	0.61	−0.02	−0.08	−0.07	9.
−0.28	−0.21	0.72	−0.23	−0.34	0.83	−0.04	−0.20	0.08	−0.17	−0.07	10.
−0.09	−0.06	−0.29	−0.30	−0.12	−0.24	−0.32	−0.19	−0.02	0.02	0.11	11.
0.28	−0.16	−0.31	−0.40	−0.17	−0.28	−0.45	−0.21	−0.02	0.02	−0.08	12.
1.0	0.57	−0.27	0.14	−0.31	−0.31	−0.08	0.02	−0.05	0.04	0.07	13.
	1.0	−0.15	0.05	−0.24	−0.20	−0.13	−0.04	−0.05	−0.01	0.08	14.
		1.0	−0.50	−0.41	0.91	0.10	−0.21	0.05	−0.10	−0.03	15.
			1.0	−0.31	−0.17	0.87	0.59	−0.11	−0.03	−0.12	16.
				1.0	−0.36	−0.36	−0.23	−0.09	013	0.06	17.
					1.0	0.04	−0.21	0.10	−0.16	−0.07	18.
						1.0	0.65	−0.06	−0.14	−0.05	19.
							1.0	0.17	0.04	−0.08	20.
								1.0	0.11	−0.01	21.
									1.0	0.21	23.
										1.0	27.

stances the levels were transgressive. Despite this variation, compound to compound, the means of the F_2 population did not differ significantly from those of the F_1's with two exceptions: *d*-limonene and unknown compound 13.

Intercorrelation and Factor Analysis

The intercorrelation coefficients are presented in Table 3, and the singles linkage diagram of factor analysis in Fig. 5. The latter allowed us to cluster groups of highly positively intercorrelated compounds and to examine them as units (factors).

Factor one

Factor 1 consists of six compounds: α- and β-pinenes, camphene, tricyclene, borneol, and an unknown bicyclic, compound 10 (Fig. 5). As a unit, it displays a remarkable degree of positive intercorrelation, with the lowest correlation coefficient between pairs 10-1 being 0.7 (Table 3). The highest correlation coefficient existed between α-pinene and β-pinene (0.9953). Structurally, Factor 1 represents the known bicyclic compounds under study. Thus, it seems apparent in view of their high degree of intercorrelation and structural similarity that Factor 1 represents a highly cohesive biosynthetic and perhaps genetic unit. A number of biosynthetic models dealing with these compounds have been formulated from a variety of experimental data (Ruzicka, 1953; Sandermann, 1962; Scora and Mann,

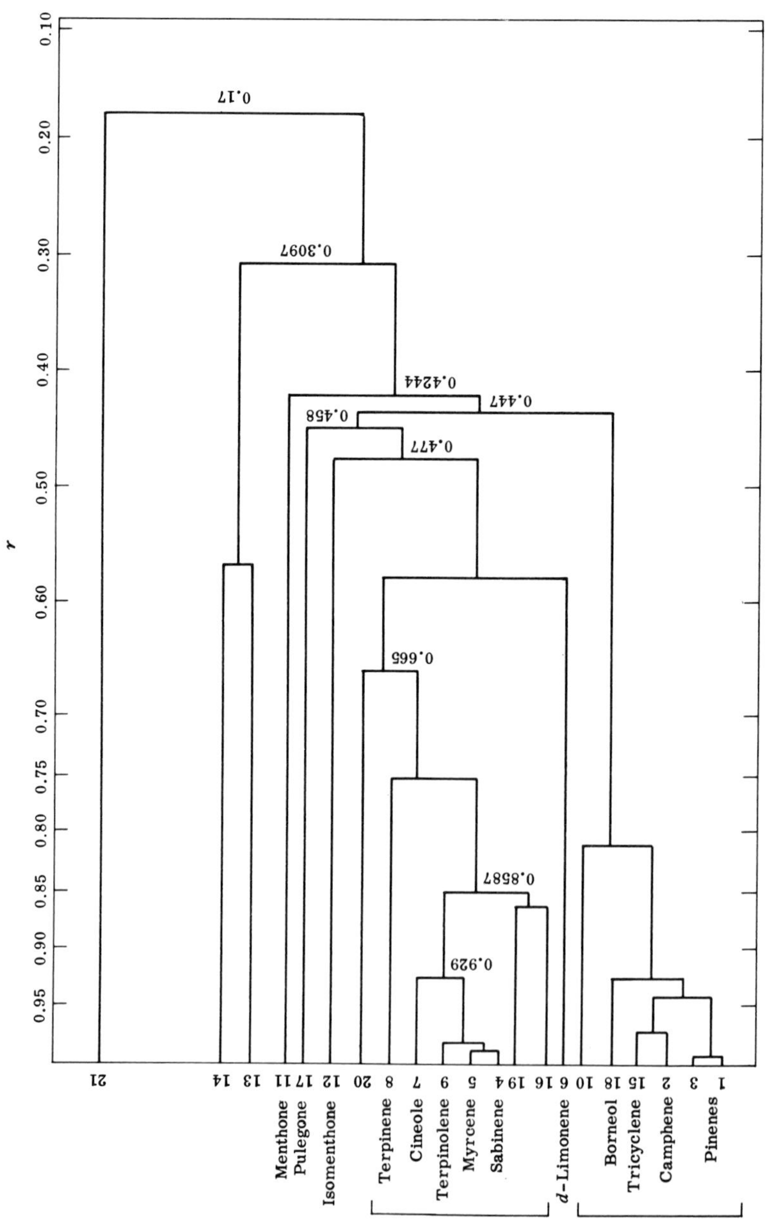

FIG. 5. A single-linkage clustering of monoterpenes in the F_2 generation by correlation coefficients.

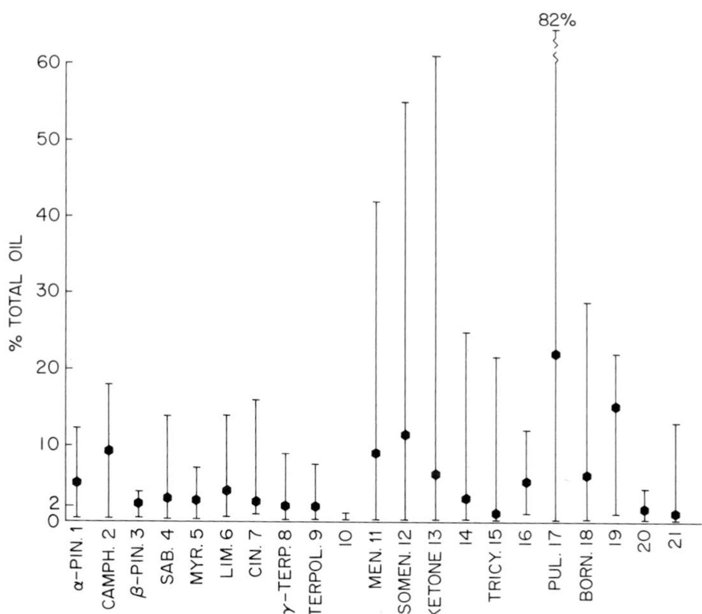

FIG. 6. A synopsis of the total monoterpene variation in the F_2 generation. Height of bar represents the range; hexagon, the mean (x). For names of compounds, see corresponding numbers in Table 3.

1967; Zavarin, 1970; Zavarin and Cobb, 1970). Most of these schemes have biosynthetically arranged the bicyclics around a common intermediate. The interrelationships uncovered in our data would further support such a model. We would modify it, however, to bring the pinenes into closer biosynthetic unity with borneol, camphene, and tricyclene (Fig. 7). This common intermediate theory for Factor 1 would nicely account for the positive intercorrelations and for the genetic patterns discussed in the next section, but its existence is by no means firmly established. What is actually functioning as the intermediate, if any, is in no way determinable from our data, and the carbonium ion is used in ignorance. Indeed, the presence or absence of a common intermediate may not be as of much importance as unique enzymes which can operate on a ubiquitous intermediate.

Factor Two

A second unit of highly positively intercorrelated compounds was revealed through factor analysis: myrcene, sabinene, terpinolene, terpinene, unknowns 16 and 19, and 1,8-cineole (Fig. 5). The structurally unlike pair,

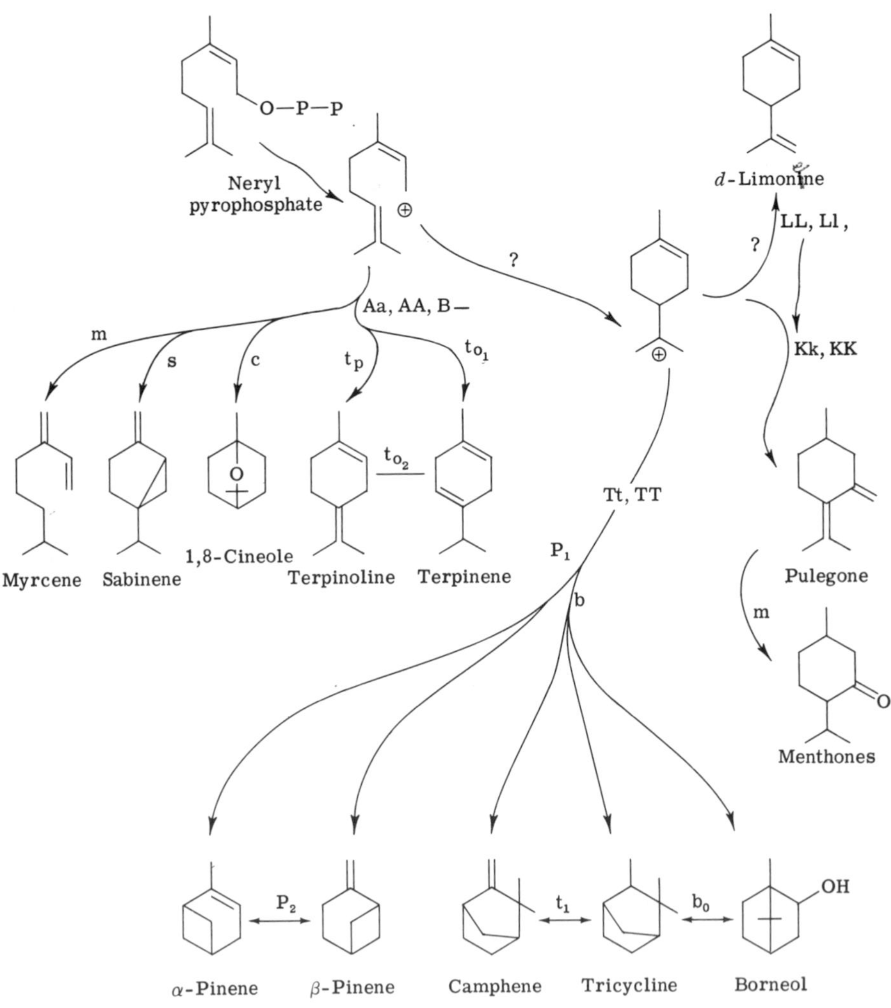

FIG. 7. Biosynthetic and genetic synopsis of the monoterpenes in the *Hedeoma drummondii* complex. Lower case letters represent alternative sites of gene and/or enzymatic action.

sabinene and myrcene, displayed the highest correlation coefficient for this unit, 0.9906 (Table 3) and is at variance with previous data (Zavarin *et al.*, 1971). Although Factor 2 is structurally more diverse than Factor 1, its strong cohesiveness again suggests biosynthetic and/or genetic unity. Indeed, it is interesting to note that two F_2's were quantitatively almost exclusively Factor 2. Again with the high level of positive correlation, it is tempting to model these compounds through a common intermediate. Yet, because of their structural diversity, such a model must be presented with even greater caution. It is conceivable that their unity is the result of diverse but tightly linked enzymes which operate independently (Fig. 7).

Since an orthogonal rotation (verimax) was used to generate Factors 1 and 2, we know that these factors represent independent modes of variation. A comparison of the intercorrelations (Table 3) shows that no compound that is highly loaded on Factor 1 is well correlated with those compounds highly loaded on Factor 2. Interestingly, both Factors, 1 and 2, are the principal constituents of the parental species *H. reverchonii* var. *serpyllifolium*.

d-Limonene

d-Limonene is a major constituent of both *H. r.* var. *serpyllifolium* and *H. drummondii* and displays a number of weak positive correlations with members of Factor 1 (0.6) and negative correlations with Factor 2 (Table 3). Within the *H. drummondii* system, *d*-limonene is highly negatively correlated with isomenthone and other ketones through developmental time (Firmage, 1971). Thus, although not readily apparent in our correlation data, *d*-limonene may be an integral part of the biosynthetic schemes for ketones.

Pulegone, Menthone, and Isomenthone

The three ketones pulegone, menthone, and isomenthone are the major terpene constituents of the *H. drummondii* parent. They display no significant positive correlations either internally or to other monoterpene products. Indeed, if any trend in correlation exists it is a negative one. This is most pronounced with the hydrocarbon fractions (Table 3) and weakly reiterates what has been said by a number of researchers (see Loomis, 1969; Firmage and Irving, 1971), that the oxygenated fraction is made in competition with, or at the expense of, hydrocarbons. Because of their ubiquity a number of biosynthetic schemes have been proposed for these monocyclic ketones (Loomis, 1969; Scora and Mann, 1967). In all, a sequential reduction series is envisioned usually beginning with piperitenone and terminating with menthone or isomenthone (Fig. 7). With such a sequence, negative correlations would be anticipated (Table 3). It also might be anticipated that these would be relatively low, as the system would be an open and reversible one.

Correlations with Morphology and Fertility

The correlation between the terpenes and morphological attributes have interesting systematic implications. None of the 21 monoterpenes had any significant positive or negative correlation to any single or group of morphological characters (Table 3). Thus, monoterpenes and morphology, in this complex of species, will segregate independently in F_2 generations. It is quite conceivable that in natural variation one could obtain, through hybridization, plants morphologically of one species and chemically of another, and indeed this is realized in the field variations encountered in the central Texas populations.

Likewise, fertility (0–100 percent in the F_2) showed no significant correlations; thus, reproductive success of F_2 recombinants (as measured by fertility) would be independent of their terpenoid products (Table 3).

Genetic Analysis of Monoterpenes

In an attempt to examine the genetic component of the monoterpene patterns, frequency distributions were plotted for each of the compounds examined in the F_2 generation. The basic premise is that if a large number

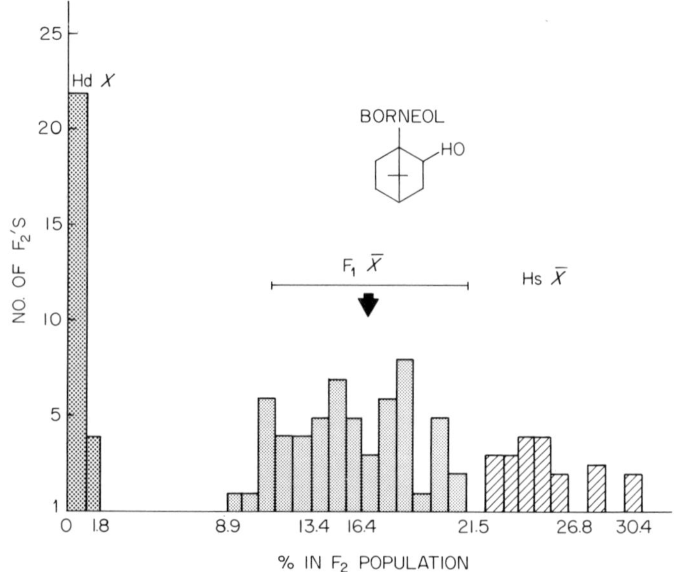

FIG. 8. Frequency distribution of borneol in F_2 generation, as representative of Factor 1 compounds.

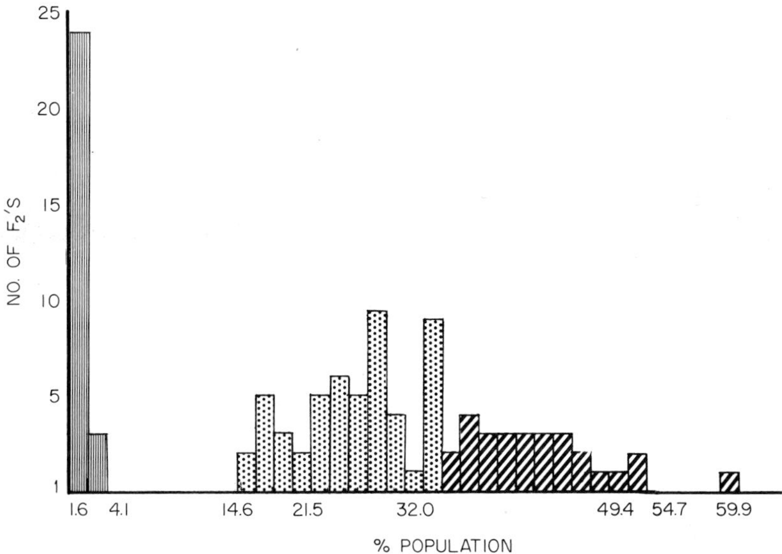

FIG. 9. Frequency distribution of an aggregation of all Factor 1 (bicyclic) compounds: α- and β-pinene, tricyclene, borneol, compound 10, camphene.

of genes govern the quantitative state of a single monoterpene we would expect the frequency distribution to approximate a Gaussian curve, whereas with one or few closely linked genes we would expect a frequency distribution with a high degree of modality. As the parental strains used in this work were highly inbred, and a high degree of initial homozygosity insured, we could also examine for Mendelian and neo-Mendelian ratios.

FACTOR ONE

The first group studied was the bicyclic compounds of Factor 1: α- and β-pinenes, camphene, tricyclene, borneol, unknown compound 10. As would be suggested by their high degree of positive intercorrelation, these compounds are unified genetically; that is, they display nearly identical frequency diagrams and can be treated as a unit. Using borneol and the bicyclic composite as examples, the levels of the two parents fell, as expected, on opposite ends of the frequency distribution and the range and mean of the F_1's at the mid-values (Figs. 8 and 9). Throughout Factor 1 there was a strong bimodality and a suggestion of trimodality (Fig. 9). The numbers of individuals filling the two major modes (low levels and medium plus high levels) have a nearly perfect 1:3 ratio to each other. The second mode (medium and high levels) can be further subdivided into two submodes in an approximate 2:1 ratio. This is further shown by the data for α-pinene (Fig. 10, Table 4) where there is a strong suggestion of a single gene model of inheritance segregating in classic Mendelian fashion, 1:2:1. Genotypes can be tentative assigned: TT, *H. r.* var. *serpyllifolium*; Tt, F_1 hybrid; tt, *H. drummondii* (Figs. 7 and 10).

As Factor One segregates as a highly intercorrelated unit rather than as individual compounds, the Mendelian pattern of inheritance more likely mirrors the inheritance of a common precursor rather than individual compounds. We could thus envision a single gene or group of tightly linked genes governing this intermediate and the production of individual compounds mediated by nonspecific enzyme steps (Fig. 7). This is, however, only one model out of perhaps several. For example, one could also suggest with the data at hand, a series of linked genes for specific enzymes which segregate and act in concert (Fig. 7). Moreover, as the correlation of Factor 1 compounds was not perfect and each compound was occasionally transgressive over the parent and F_1, a genetic model will necessarily need to include epistasis and/or modifying genes to account fully for the quantitative varition of these compounds in the F_2 generation.

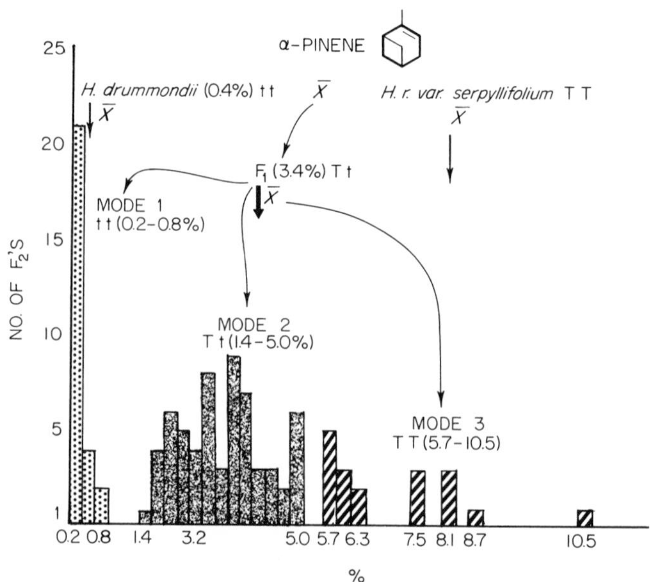

Fig. 10. Frequency of α-pinene in F_2 generation and genetic model of bicyclic inheritance. T is dominant to t.

TABLE 4

CHI-SQUARE ANALYSIS OF THE F_2 SEGREGATION OF α-PINENE, REPRESENTATIVE OF FACTOR 1

Mode	No.	Expected ratio	Actual ratio	X^2
1	27	1	1.0	
2	61	2	2.3	$P = 0.14$
3	18	1	0.7	
2 + 3	79	3	3.0	$P = 1.0$

FACTOR TWO

Myrcene, sabinene, 1,8-cineole, terpinene, terpinolene, and unknowns 16 and 19, are diverse monoterpenes but display a unified inheritance pattern. Thus we are again possibly monitoring the genetics of a common precursor. The frequency distribution of sabinene (Fig. 11) is representative of this group. The distribution is possibly pentamodal. Its inheritance, then, is not that of single gene action, but neither is it that of a large number of genes. A number of genetic models could be developed for Factor 2 using two to four linked or nonlinked genes. Such models would necessarily have to satisfy a number of conditions, however: (1) all compounds in Factor 2 are in at least one or two F_2 plants transgressive to parental variation; (2) the F_2's have a narrow range of variance suggesting a uniform genotype; (3) the parents because of their inbreeding and production of a uniform F_1 should be homozygous for all loci.

A tentative and simple model using two diallelic genes A,a, and B,b can be formulated (Fig. 11). These would be located on separate chromosomes, with A slightly epistatic to B. *H. drummondii* would be represented in the model as aaBB and would produce low quantities of the Factor 2 compounds through the absence of the A allele. *H. r. serpyllifolium* would also be homozygous, AAbb, but would produce significant quantities through AA. A cross of these two homozygous genotypes would produce a genetically uniform F_1, AaBb. Selfing of the latter would yield nine different genotypes (a dihybrid cross) which could be grouped into the five modes of the F_2 frequency distribution and in the approximate ratios (Fig. 11). This array of genotypes in the F_2 would show an increasing production of sabinene (or other Factor 2 compound) with increasing proportions of A, primarily, and B, secondarily, in the genotype. Thus, AABb and AABB would represent the transgressive F_2 genotypes. The expected ratio of such a

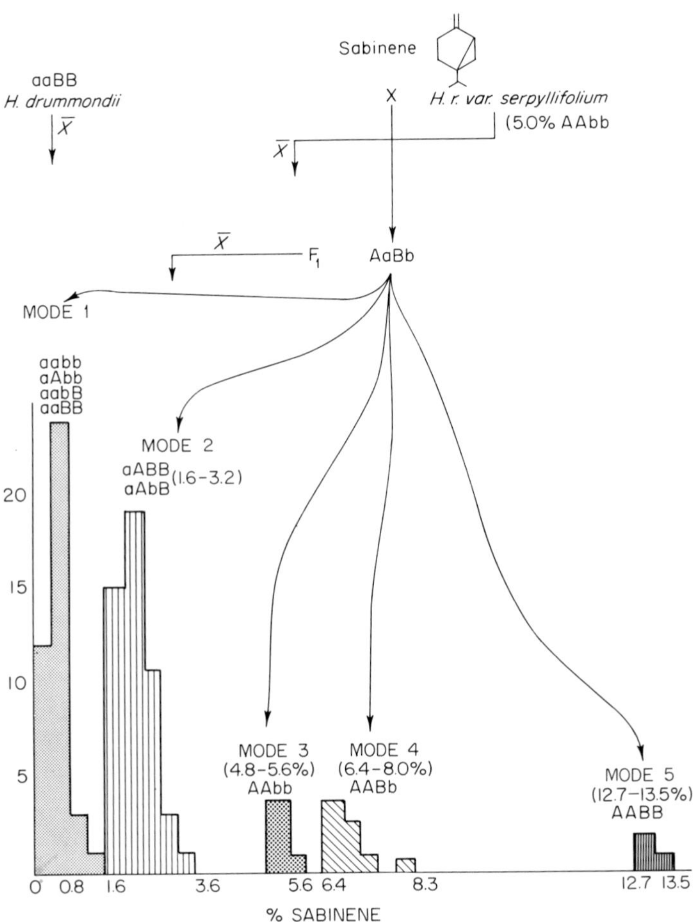

FIG. 11. Frequency of sabinene in F_2 generation and genetic model of inheritance for Factor 2. Model is based on two genes (A,B) with two alleles (A, a and B, b); these are located on separate chromosomes with A slightly epistatic to B. Quantitative level of Factor 2 compound increases with increasing proportion of A in genome (see text).

model, 6:6:1:2:1, approximates the actual ratio obtained (Table 5). No doubt, a better fit could be obtained using more complicated models, but in view of the evidence, further statistical gamesmanship seems unprofitable.

d-LIMONENE

Despite *d*-limonene's seeming importance in terpene biosynthesis, little can be said of its genetic base from our data. The parents showed no sig-

TABLE 5

CHI-SQUARE ANALYSIS OF THE F_2 SEGREGATION OF SABINENE, REPRESENTATIVE OF FACTOR 2

Mode	No.	Expected ratio	Actual ratio	X^2
1 (0–1.2)	40	6	6.1	
2 (1.6–3.2)	49	6	7.4	
3 (4.8–5.6)	5	1	0.8	$P = 0.25$
4 (6.4–8.0)	9	2	1.4	
5 (12.7–13.5)	3	1	0.5	

nificant difference in quantitative levels (3.6 and 4.5 percent). The F_1 and F_2 populations displayed widely varying levels of *d*-limonene which were for the most part transgressive. The F_2 frequency distribution (Fig. 12) offered little suggestion of the *d*-limonene's inheritance save that is perhaps ogligogenic. However, much of the F_2 variation might be due to varying levels of biosynthetic competition as it is perhaps a key compound in monoterpene interconversions.

PULEGONE, ISOMENTHONE, AND MENTHONE

Individually these ketones do not display interpretable frequency patterns, at least, in Mendelian terms. If they are treated as a unit, however

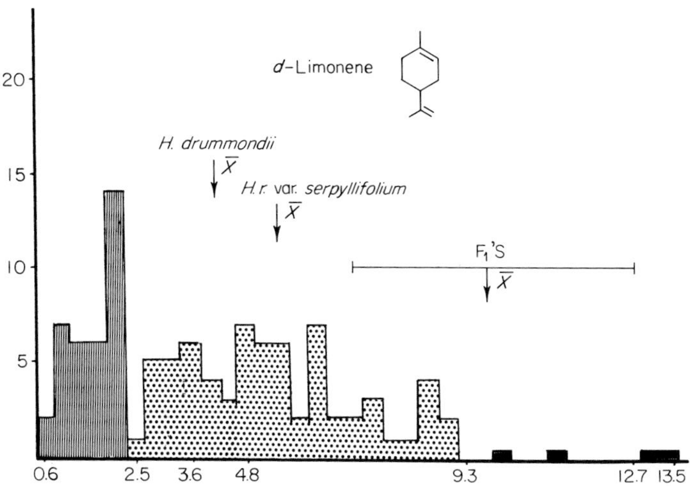

FIG. 12. Frequency of *d*-limonene in F_2 generation.

(Fig. 13), a classic 1:2:1 ratio emerges (Table 6), which suggests one or few tightly linked genes segregating as a Mendelian unit. Using such a model, genotypes could be as follows: kk, *H. r. serpyllifolium*; Kk, F_1, and KK, *H. drummondii*. These alleles would presumably govern the level of a cyclic precursor which would move through a reduction series (Fig. 7).

A Statistical Estimate of Gene Number

Although frequency plots of large controlled F_2 generations are a standard analytical tool for determining genetic background, the minimal num-

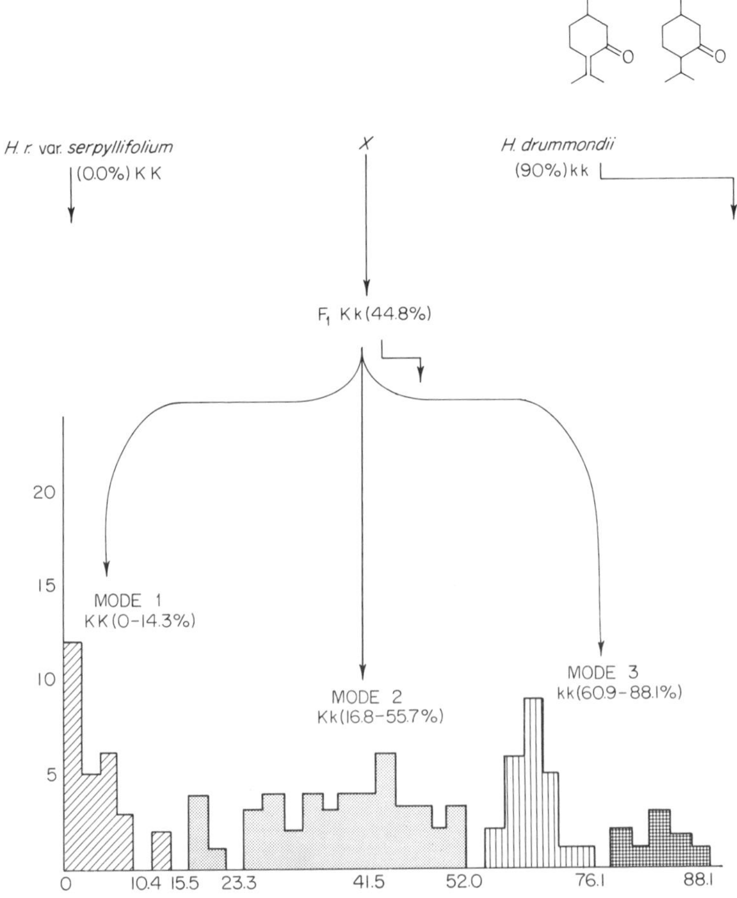

Fig. 13. Frequency distribution of pulegone, menthone, and isomenthone in F_2 generation and accorded genetic model based on one gene with two alleles.

TABLE 6

CHI-SQUARE ANALYSIS OF THE F_2 SEGREGATION OF PULEGONE, ISOMENTHONE, AND MENTHONE

Mode	No.	Expected ratio	Actual ratio	X^2
1	28	1	1.0	
2	45	2	1.8	$P = 0.24$
3	33	1	1.2	
2 + 3	78	3	2.8	$P = 0.9$

ber of genes can also be determined using the formula developed by Sewall Wright (1934) (see also Charles and Goodwin, 1953):

$$N = \frac{(V_{p2} - V_{p1})^2}{8(\sigma_2^2 - \sigma_1^2)}$$

TABLE 7

ESTIMATE OF MINIMAL NUMBER OF GENES

Compound	N
α-Pinene	2
Camphene	3
β-Pinene	2
Sabinene	1
Myrcene	1
d-Limonene	1
1,8-Cineole	4
Terpinene	1
Terpinoline	1
Compound 10	4
Menthone	1
Isomenthone	7
Compound 13 ketone	1
Compound 14	1
Tricyclene	3
Compound 16	1
Pulegone	1
Borneol	2
Compound 19	1
Compound 20	1
Compound 21	0

In the formula, V_{p1} and V_{p2} represent the average of the two parents with respect to a monoterpene character and σ_2, σ_1 the standard deviations of the F_2 and F_1. The validity of N, the minimal number of genes, depends on the number of conditions which must be met by the experimental breeding program (see Charles and Goodwin, 1953).

Making these calculations for the 21 terpenes (Table 7), we have an approximate estimate of minimal genes within the limitations of the formula. In general the results indicate that most terpenes are controlled by a small number of genes. In the specifics there is, however, variance between the two approaches. The strength of Wright's formulation depends, for the most part, on large parental and F_1 sample sizes. In our program we are not able to provide this prerequisite adequately, and these minimal estimates must be treated as only approximations.

Discussion

The problems in the genetics and biosynthesis of monoterpenes are far from solved. In *Hedeoma*, and possibly in other systems, a substantial portion of their biogenetic base is seemingly simplistic, involving a few major genes (Fig. 7). This is in keeping with inferences from several researches (Forde, 1964; Hanover, 1966b; Zavarin *et al.*, 1969). Our data suggest further simplification, but we must proceed with caution in this reduction of complexity; although having their uncomplicated facets, biological systems, as dynamic wholes, may be exceedingly complex. Pervading our own data are indications of more complex levels of gene and enzymatic action. Epistasis, modifying genes, and enzyme complementation, to which we have alluded, are all characteristic of multigenic systems. Because of this, in our biogenetic synopsis (Fig. 7) a number of additional and/or alternative sites of gene action have been designated.

Moreover, to evaluate properly genetics through breeding, the organism's limitations must be realized and the nature of the character studied must be understood. Patterns of variation in an F_2 generation take on genetic significance only when there is a cellular environment conducive to free recombination. That is, there must be a substantial degree of normal chromosomal pairing and chromatid exchange in the F_1 parent. In our work, although the F_1 had a lowered fertility, there was a high degree of normal pairing as well as a number of meiotic irregularities (univalent formation). There is then the indication of a partial suppression of recombination, and our conclusions must be tempered accordingly.

To understand the monoterpene as a character, the gap between the gene and its expression must be closed. There must be greater certainty in the proposed series of interconversions and enzymatics. Lacking this, our genetic and biogenetic models can only stand as guides.

The systematic implications of our data are of considerable importance. Ideally, systematists value those chemical characters which, through their biosynthetic and genetic complexity, can be used to delineate clearly reproductive barriers as well as to provide ample variation with gene flow. The monoterpenes of the *H. drummondii* group, through their perhaps deceptive simplicity, can only partially fulfill these goals. Considering the F_2 segregation, complex patterns of natural interbreeding could only be partly detected by monoterpene analysis. To delimit species by employing these constituents, and their proposed genetic base, would also be dubious. Nevertheless, one of the most important findings in our experimental work was the independent segregation of morphology and terpenes. Thus, an excellent mirror of gene flow would be gained utilizing monoterpene diversity *and* morphology. Moreover, if we are indeed, with monoterpenes, investigating a relatively uncomplicated biogenetic system, it seems that the hope of using biochemical pathways as evolutionary guideposts may well be within our grasp.

Acknowledgments

This work was supported by a National Science Foundation Grant (GB-12910) awarded to the senior author; computer time was furnished by Colorado State University. We wish also to express our sincere thanks to Mrs. Caryn Stone for her technical assistance and patience. Thanks also go to Drs. David V. Clark, B. L. Turner, E. von Rudloff, and Victor C. Runeckles for their helpful discussions, suggestions, and support. We are grateful to others who assisted in many ways: Mr. David Firmage for his laboratory and field work, Linda Nimmo and Evalene Wyatt for their assistance in the preparation of the manuscript, and, to the senior author, to Jody Lubrecht, and to Joellen Irving for their support and encouragement.

References

Adams, R. P. 1970. *Phytochemistry* **9**:397–402.

Alston, R. E., and B. L. Turner. 1963a. "Biochemical Systematics." Prentice-Hall, Englewood Cliffs, New Jersey.

Alston, R. E., and B. L. Turner. 1963b. *Amer. J. Bot.* **50**:159–173.

Alston, R. E., H. Rosler, K. Naifeh, and T. J. Mabry. 1965. *Proc. Nat. Acad. Sci. U.S.* **54**:1458–1465.

Battu, R. G., and H. W. Youngken, Jr. 1966. *Lloydia* **29**:360–367.

Burbott, A. J., and W. D. Loomis. 1967. *Plant Physiol.* **42**:20.

Charles, D. R., and R. H. Goodwin. 1953. *Amer. Natur.* **77**:53–69.

Critchfield, W. B. 1966. *U.S. Forest Serv., Res. Pap.* **NC–6**:34–44.

Critchfield, W. B. 1967. *Silvae Genet.* **16**:89–97.

DeWet, J. M. J. 1967. *Amer. J. Bot.* **54**:384–387.

DeWet, J. M. J., and B. D. Scott. 1965. *Bot. Gaz. (Chicago)* **126**:209–214.

Emboden, W. A., and H. Lewis. 1967. *Brittonia* **19**:152–160.
Firmage, D. 1971. Unpublished observations, Univ. of Montana, Missoula, Montana.
Firmage, D., and R. S. Irving. 1971. *Annu. Phytochem. Soc. Meet., 11th, Monterrey, Mexico.*
Forde, M. B. 1964. *N. Z. J. Bot.* **2**(1): 53–59.
Fujita, Y. 1965a. *Shokubutsugaku Zasshi* **78**(924):212–219.
Fujita, Y. 1965b. *Shokubutsugaku Zasshi* **78**(925):245–252.
Fujita, Y. 1965c. *Osaka Kogyo Gijutsu Shikenjo Hokoku* No. 306–2.
Habeck, J. R., and T. W. Weaver. 1969. *Can. J. Bot.* **47**:1565–1570.
Hanover, J. W. 1966a. *Heredity* **21**:73–84.
Hanover, J. W. 1966b. *Forest Sci.* **12**:447–450.
Irving, R. S. 1968. Ph.D. Thesis, Univ. of Texas, Austin, Texas.
Kremers, R. E. 1922. *J. Biol. Chem.* **50**:31.
Loomis, W. D. 1969. *In* "Terpenoids in Plants" (J. B. Pridham, ed.). Academic Press, New York.
Loomis, W. D. 1972. *In* "Terpenoids: Structure, Function and Biogenesis" (V. C. Runeckles and T. J. Mabry, eds.). Academic Press, New York.
Mirov, J. T. 1956. *Can. J. Bot.* **34**:443–457.
Murray, M. J. 1960a. *Genetics* **45**:925–929.
Murray, M. J. 1960b. *Genetics* **45**:931–937.
Reitsema, R. H. 1958. *J. Amer. Pharm. Ass.* **47**:267–269.
Ruzicka, L. 1953. *Experientia* **9**:357–396.
Sandermann, W. 1962. *Holzforschung* **16**:65–74.
Scora, R. W. 1967a. *Amer. J. Bot.* **54**:446–452.
Scora, R. W. 1967b. *Taxon* **16**:499–505.
Scora, R. W. 1970. *Taxon* **19**:215–228.
Scora, R. W., and J. D. Mann. 1967. *Lloydia* **30**:236–241.
Smith, D. M., and D. A. Levin. 1963. *Amer. J. Bot.* **50**:952–958.
Stone, D. E. 1964. *Amer. J. Bot.* **51**:687–688.
Stone, D. E., G. A. Adrouny, and S. Adrouny. 1965. *Brittonia* **17**:97–106.
Stone, D. E., G. A. Adrouny, and R. H. Flake. 1969. *Amer. J. Bot.* **56**:928–235.
Turner, B. L. 1967. *Pure Appl. Chem.* **14**:189–213.
Turner, B. L. 1969. *Taxon* **18**:134–151.
Wright, S. 1934. *Genetics* **19**:537–551.
Zavarin, E. 1970. *Phytochemistry* **9**:1049–1063.
Zavarin, E., and F. W. Cobb, Jr. 1970. *Phytochemistry* **9**:2509–2515.
Zavarin, E., W. B. Critchfield, and K. Snajberk. 1969. *Can. J. Bot.* **47**:1443–1453.
Zavarin, E., L. Lawrence, and M. C. Thomas. 1971. *Phytochemistry* **10**:379–393.

CONFIRMATION OF A CLINAL PATTERN OF CHEMICAL DIFFERENTIATION IN *Juniperus virginiana* FROM TERPENOID DATA OBTAINED IN SUCCESSIVE YEARS

ROBERT H. FLAKE,* ERNST VON RUDLOFF,† and B. L. TURNER‡

* *Electrical Engineering Department and* ‡ *Botany Department, University of Texas, Austin, Texas, and* † *Prairie Research Laboratory, National Research Council of Canada, Saskatoon, Canada*

Introduction

Juniperus virginiana (Eastern red cedar) is a widespread, weedy tree-species occurring throughout most of the eastern United States from

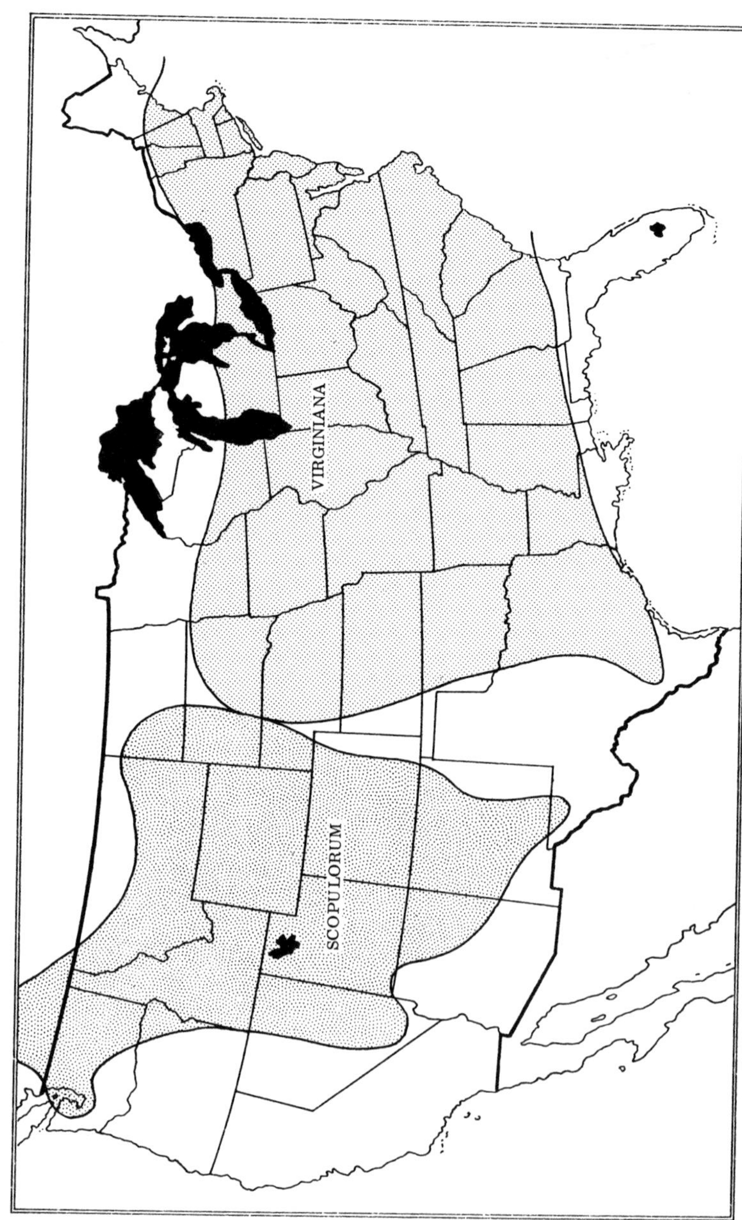

Fig. 1. General distribution of the allopatric species, *Juniperus virginiana* and *J. scopulorum*.

Canada to near the Mexican border (Fig. 1). It is readily recognized by its dark green, dense foliage, especially during the winter months, for it most often occurs in association with winter-deciduous hardwoods, usually in sandy soils in the western part of its range; in the more eastern, wetter portions of its range, however, it occurs in a wide variety of habitat types, often being the sole dominant on shallow limestone soils.

The species is primarily allopatric but areas of sympatry occur with at least three species, *Juniperus horizontalis*, *J. scopulorum*, and *J. ashei* (Little, 1971), and several authors (Fassett, 1945; Hall, 1952; Van Haverbeke, 1968) have reported the occurrence of putative hybrid swarms of each of these species with *J. virginiana* in such areas of sympatry. Perhaps the most thoroughly studied of these is that involving allopatric introgression (hybridization with backcrossing among allopatric taxa so that gene flow over an extensive region is permitted) between *J. virginiana* and *J. ashei*.

In a series of papers beginning in 1952, Hall (Hall, 1952, 1955; Hall and Carr, 1962) has maintained that these two, very different species freely exchange genes in central Texas, east-central Oklahoma and north-central Arkansas so that genes of *J. ashei* "appear in red cedar (*J. virginiana*) populations of the granite outcrops along the fall line of the Piedmont upland in Georgia" (Hall and Carr, 1962). In fact, the study of hybridization between these two species has been repeatedly referred to as one of the most detailed, best documented examples of allopatric introgression in the literature (Anderson, 1953; Grant, 1958; Davis and Heywood, 1963).

Our interest in this problem began in 1967. We set out, *not* to test the validity of the introgression hypothesis, but rather to use the sensitivity of gas–liquid chromatography (GLC) to detect the direction and extent of gene flow from *J. ashei* into the populations of *J. virginiana*. In short, we accepted that introgression between these two species was occurring.

Because of our original bias in favor of the hypothesis for introgression, it seems appropriate to note here that in order to avoid any bias in our sample data we coded all the material analyzed by one of us (E.v.R.), without indicating any identification. After making several population samples for terpene analysis from sites where the two species occurred alone, and after ascertaining that the two species were readily distinguished both quantitatively and qualitatively (Vinutha and Von Rudloff, 1968), we then proceeded to sample the putative hybrid populations of these species. To make a long and disheartening story short, we were unable to find a single F_1 hybrid or backcross between these two species, even at localities where the taxa occur side by side, including the very populations upon which Hall (1955), in part, based his hypothesis for introgression (Von Rudloff *et al.*, 1967; Turner, 1970; Irving and Turner, 1973).

Since introgression as a source of the rather obvious regional variation in the morphology (mainly habit) of *Juniperus virginiana* was negated by the above studies, we turned our attention to the question of clinal variation in *J. virginiana* itself, taking our clue from the observations of Barber and Jackson (1957), who suggested that the variation found in the latter species was due to differential selection along an ecological gradient. The so-called introgressed populations of *J. virginiana* (consisting of trees with wider aprons, shorter terminal shoots, etc.) were believed to be due to selective influences working on a variable gene pool in the western part of its range so as to produce populations superficially like *J. ashei*. In short, these authors suggested that the morphological variation found in *J. virginiana* was clinal, i.e., the species has formed, or is in the process of forming, regional races as a result of adaptational mechanisms arising out of its own gene pool, this being unrelated to the possible influx of genes from the largely allopatric *J. ashei*.

Subsequent analysis of *Juniperus virginiana* populations along a 1500-mile transect from Texas to Washington, D. C. in the winter of 1968, using the same coding procedures referred to above, showed that the relatively vague clinal variation in habit was paralleled by a rather distinct clinal variation in the volatile components (Flake *et al.*, 1969). In fact, we were sufficiently impressed by the results that we felt compelled to resample the populations a second time for verification. That is, the relatively few individuals collected per populations during the first year (only 10 plants at each of nine sites) might have given us a misleading picture of the populational structuring in *J. virginiana*. Because of this, we sampled the same route a year later, with double the population sample size of the original study. We also extended the transect an additional 150 miles by adding a population to the Texas end of the transect. It is the purpose of this paper, then, to compare the results obtained from these two studies.

Population Data

The *J. virginiana* data were collected in December, 1967, and January, 1968, and resampled again in January, 1969. Foliage from ten plants at nine population sites was sampled the first winter at approximately 150-mile intervals along a transect from the northeastern border of Texas to Washington, D. C., together with two additional populations sampled near Durham, N. C., and South Hill, Virginia.

The populations along the transect were resampled a year after the first study, along with one additional population, extending the transect approximately 150 miles deeper into Texas. Samples from twenty trees were

TABLE 1

COLLECTION DATA FOR THE POPULATIONS INDICATED IN FIGURES 2 AND 3

Population No.	Locality	Collection dates	
0	Texas: Milam County	—	Jan. 22, 1969
1	Texas: Panola County	Jan. 19, 1968	Jan. 22, 1969
2	Louisiana: Claiborne Parish	Jan. 19, 1968	Jan. 23, 1969
3	Mississippi:Leflore County	Jan. 20, 1968	Jan. 23, 1969
4	Mississippi:Monroe County	Jan. 20, 1968	Jan. 23, 1969
5	Alabama: Lawrence County	Jan. 20, 1968	Jan. 23, 1969
6	Georgia: Dade County	Jan. 21, 1968	Jan. 24, 1969
7	North Carolina: Catawba County	Jan. 22, 1968	Jan. 24, 1969
8	North Carolina: Alamance County	Jan. 22, 1968	Jan. 25, 1969
9	Maryland: Prince George County	Jan. 25, 1968	Jan. 25, 1969
10	North Carolina: Durham County	Dec. 8, 1967	—
11	Virginia: Mecklenburg County	Dec. 9, 1967	—

collected at each site the second year, doubling the sample size from each population over that of the previous year. Sample site localities and collection dates are listed in Table 1. Voucher material for each of the populations sampled are deposited at the University of Texas Herbarium.

Chemical Analysis

A gas chromatographic analysis of the terpenoid oil obtained by steam distillation of the *J. virginiana* foliage yielded percentage distributions of 37 distinct compounds that were detected in sufficient quantities to be well above instrumental error as well as having the appearance of being present in reasonably consistent amounts within the individual populations.

The chemical analysis was described in detail previously (Von Rudloff, 1968; Vinutha and Von Rudloff, 1968). The individual components, 1 to 37, used in the analysis were α-pinene and thujene (combined), sabinene, 3-carene, myrcene, limonene, β-phellandrene,* γ-terpinene, terpinolene, "peak 16" (Vinutha and Von Rudloff, 1968), "peak 17," linalool, methyl citronellate, "peak 22," 4-terpinenol, "peak 25," estragole, "peak 27," methyl vinyl anisole, δ-cadinene, "peak 30," citronellol, "peak 32," nerol, "peak 34," geraniol, "peak 37," "peak 39," methyl eugenol, "peak 41 and

* Limonene and β-phellandrene were difficult to separate and were therefore combined in the analysis of the second-year data.

42," elemol, elemyl acetate, "peak 46," γ-eudesmol, "peak 49," α-eudesmol and elemicin (combined), β-eudesmol, "peak 53," and acetate II.

Characterization

The terpenoid patterns from foliage samples of individual plants representing each population sample were averaged. The terpenoids have different ranges of variation and do not vary independently. Thus, the populations are represented by vectors in a character space in which a "feature weighting metric" is selected to account for the differences in variability of the chemical characters. Uncorrelated, standardized characters are first constructed from a principal components analysis transformation applied to the data which are thereby standardized. For each character a weighting coefficient that reflects the influence of local selection pressures is then computed from the estimated population variance of the character. The taxonomic distance for a pair of populations indexed j and k is defined as

$$d(j, k) = \left(\sum_{n=1}^{N} \omega_n{}^2 (X_{nj} - X_{nk})^2 \right)^{1/2}$$

where N is the number of transformed characters considered, and X_{nj} is the average value of the nth uncorrelated character from the sample of the jth population. The weighting coefficients, ω_n, $n = 1, 2, \ldots, N$, satisfy the constraint

$$\sum_{n=1}^{N} \omega_n{}^2 = 1$$

The measure of similarity for populations j and k is defined as

$$S(j, k) = 1 - c \cdot d(j, k)$$

where the constant c is conveniently chosen to be 0.1 to ensure that $0 \leq S(j, k) \leq 1$, for all object pairs under consideration. (A measure of similarity is introduced only because the computer software that has been developed in these studies requires a matrix of similarity values rather than a matrix of taxonomic distances.)

The weighting coefficients in the taxonomic distance expression are determined from the average of the individual population sample standard deviation estimates of each uncorrelated character and the estimated pooled sample standard deviations of each character. The weighting coefficient for character n is defined as

$$\omega_n(s_{pn}, s_n) = \begin{cases} (1 - s_{pn}/s_n), & \text{if } s_{pn} \leq s_n \\ 0, & \text{if } s_{pn} > s_n \end{cases}$$

where s_{pn} is the average of the population sample estimates of the standard deviations for character n, and s_n is the pooled sample standard deviation estimate for the nth character.

The biological significance of these weighting coefficients is that they give relatively large fractional weight to those characters that are influenced by localized natural selection pressures and hence tend to have small variations on the average within each population sample, while simultaneously having different population means which results in relatively large pooled sample variations.

Clustering Technique

A number of clustering techniques have been proposed for computer-aided systematic studies. A summary of some of the earlier techniques is contained in Sokal and Sneath (1963). Flake and Turner (1968) developed a clustering method consisting of a clump-searching hierarchical classification algorithm which was applied in this study.

A brief description of the clustering concepts is presented here. A more complete presentation including a discussion of the correspondence of these computer-aided classification concepts with taxonomic concepts in biology is contained in Flake and Turner (1968).

Consider an arbitrary space of objects X. Define a measure of similarity for each pair of elements of X. Let $\mathbf{O}$ be that class of subsets o of elements of X such that each set o contains two or more elements of X. Let $\mathbf{C}$ be the class of sets o in $\mathbf{O}$ defined by the membership function f_C (Zadeh, 1965), where $f_C(o)$ is the average of the measures of similarity of all distinct pairs of elements in o. Each set o is called a *cluster* and the numerical value $f_C(o)$ is a measure of *cluster similarity* of the cluster o. $\mathbf{C}$ is therefore a class of clusters where clusters are sets of "similar objects."

The clusters of $\mathbf{O}$ are ordered according to their cluster similarity, and a hierarchy of partitions is constructed by forming particular collections of the ordered clusters called "aggregations," which will now be defined. Let $\mathbf{O}_s = \{o_i : o_i \in \mathbf{O}, i \in I_s, f_C(o) \geq s\}$ be the sublcass $\mathbf{O}_s$ of $\mathbf{O}$ indexed by a set I_s, such that a cluster o_i is a member of $\mathbf{O}_s$ if the cluster similarity $f_C(o_i)$ is at least as high as the level s, where s is some real number in [0, 1]. Now let $I_{\mathbf{A}_s} \subset I_s$ be a subset of the indexing set for $\mathbf{O}_s$ which identifies the sets contained in a particular subclass $\mathbf{A}_s$ of $\mathbf{O}_s$ having the following restrictive properties:

$$(\bigcup_{i \epsilon I_{\mathbf{A}_s}} o_i) \cap (\bigcup_{i \epsilon I_{\mathbf{A}_s} \sim I_s} o_i) = \emptyset \tag{1}$$

For every nonempty proper subindexing set

$$J \subset I_{\mathbf{A}_s}, (\bigcup_{i \epsilon J} o_i) \cap (\bigcup_{i \epsilon I_{\mathbf{A}_s} \sim J} o_i) \neq \emptyset \tag{2}$$

Then the class $A_s = \{o_i : i \in I_{A_s}\}$ is called an *aggregation* in $\mathbf{O}_s$ at the level of similarity s. Note that the collection of all aggregations at a similarity level s forms a partition of $\mathbf{O}_s$. A hierarchy of partitions is constructed by ordering aggregations according to their similarity level s.

The hierarchy of aggregations has the property that any object (cluster element) appearing in an aggregation belongs to a cluster in the aggregation that has a cluster similarity at least as large as that of the aggregation level, and its similarity to any object not in the aggregation is less than this level. In this sense, each object in an aggregation at a given level is a member of a "natural group"; that is, "groups" of similar objects, or clusters, that are separated from or dissimilar to the other objects not contained in the aggregation.

An efficient search algorithm has been developed to yield such a hierarchy of "natural groups" of objects given an object-character data matrix. The programmed search algorithm is discussed elsewhere (Flake and Turner, 1968).

Statistical Significance of Clustering Results

A primary objective of biological classification (Mayr, 1965) is not to facilitate identification, but rather in an operational sense to achieve a highly stable or "predictive" classification that is relatively invariant when new characters or new taxa are included in the study of the preexisting classification. An initial step in the direction of developing a method for assessing the predictive value of a computer-aided systematic study of populational structure is to examine a somewhat related question, i.e., what portion of a computer-generated hierarchical arrangement of populations may reasonably be expected to be invariant to the statistical fluctuations resulting from repeated resampling? A procedure for estimating the significance of the order of aggregations in a hierarchical arrangement produced by this clustering algorithm has been developed, based on properties of the cluster and aggregation definitions. The aggregations are composed on intersecting clusters with cluster similarities greater than or equal to the aggregation level, and a cluster similarity is defined as the average of the measures of similarities of all the distinct pairs of objects contained in the cluster. The statistical significance of a computed hierarchical structure may therefore be tested by applying a multiple range test (assuming normality) to differences of cluster similarities which are computed from mean population pair similarity measures.

Results and Interpretations

Populations

The countour map in Fig. 2 shows the hierarchy of aggregations for the population data along the transect in the first-year sample that were statistically significant at the 5 percent level with Duncan's Multiple Range Test. These results indicate a clinal clustering pattern from northeast toward the southwest. The Appalachian populations (9, 8, 7, and 6) cluster homogeneously at the highest significant level of similarity, with the populations that extend southwest along the transect clustering at successively lower levels of similarity. Note that population 5 was omitted in this analysis because the estimated variance associated with this population was significantly larger than for the other populations. A more detailed discussion of these results for the initial transect samples is contained in Flake, Von Rudloff, and Turner (1969).

The analysis of the more recent data sampled the succeeding year and reported here for the first time reconfirmed the initial results. The contour map in Fig. 3 illustrates the aggregation pattern for the resampled data. The aggregation contours drawn in solid lines represent statistically significant contours found in the analysis of the first year's data that appeared again in the results of the analysis of the resampled data. Further inspection of Fig. 3 indicates that a slightly better resolution was obtained from the second year data in which the population sample size was doubled. This enhanced resolution is apparent through the emergence of three steps in the cline, located in the Appalachians, the middle of the transect, and the populations west of the Sabine River on the Texas–Louisiana border. The two populations west of the Sabine River appeared to be distinctly different from the populations which occur east of the river. In fact, in both data sets, all the populations east of the Texas–Louisiana border formed a single cluster before either of the two Texas populations joined with any of those to the east.

Biogeography

Our original results (Flake and Turner, 1968) were interpreted as being consistent with what is believed to have been the phyletic history of the extant populations that comprise the species; i.e., that *J. virginiana* was originally a species belonging to an essentially northeastern flora, having occupied the ancient land mass of Appalachia in remote times but subsequently extending its range to peripheral western areas as previously submerged land become newly exposed for occupancy (Anderson, 1953).

The more recent data (Flake *et al.*, 1969) are also open to such an interpre-

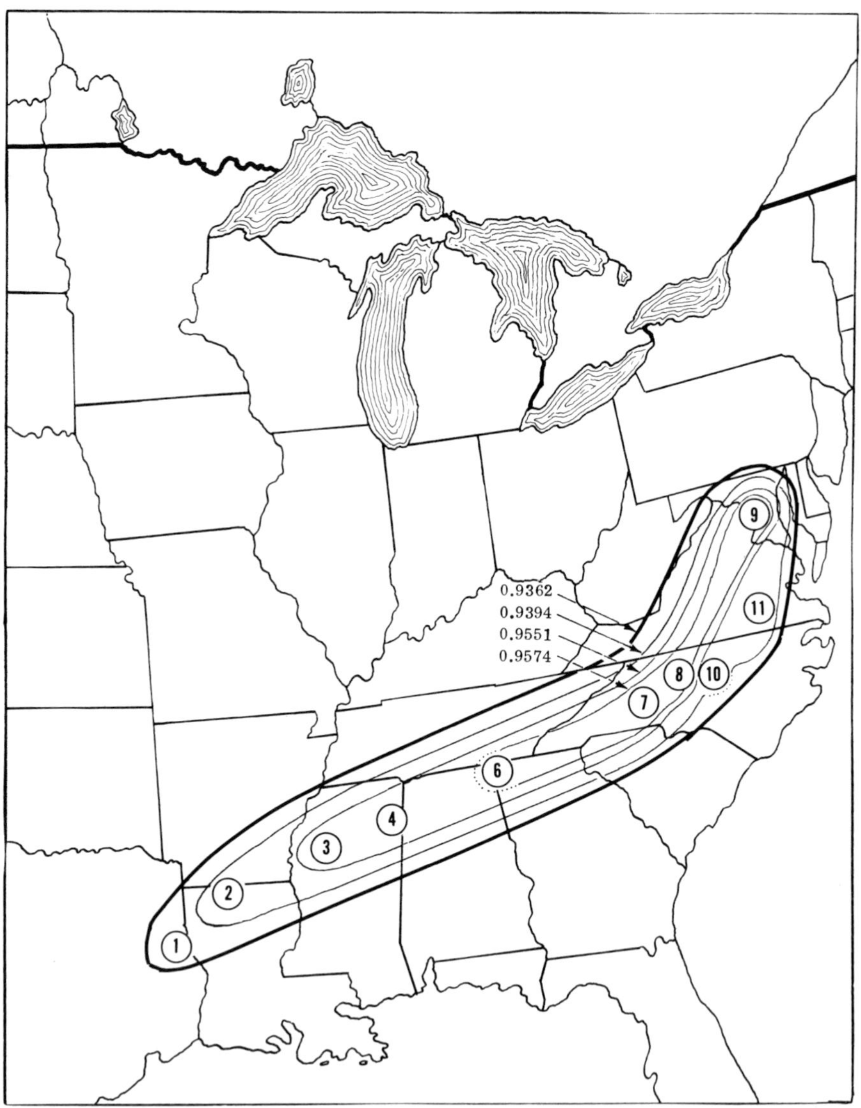

FIG. 2. Statistically significant *Juniperus virginiana* aggregation contours.

tation. However, data from the additional southwesternmost population which was added to the 1968 transect pose additional or alternative interpretations. It will be noted (Fig. 3) that the two Texas populations cluster at a relatively high level of similarity and that *all* the more eastern populations form a homogeneous cluster before either of these two populations enter the overall aggregation. This suggests that the two Texas populations are relatively remote, phyletically speaking, from the more eastern populations along the transect and thus might have had some long-time isolation in Texas, or else have been derived from populations somewhat different from those of the more eastern regions. At least, it seems unlikely that the

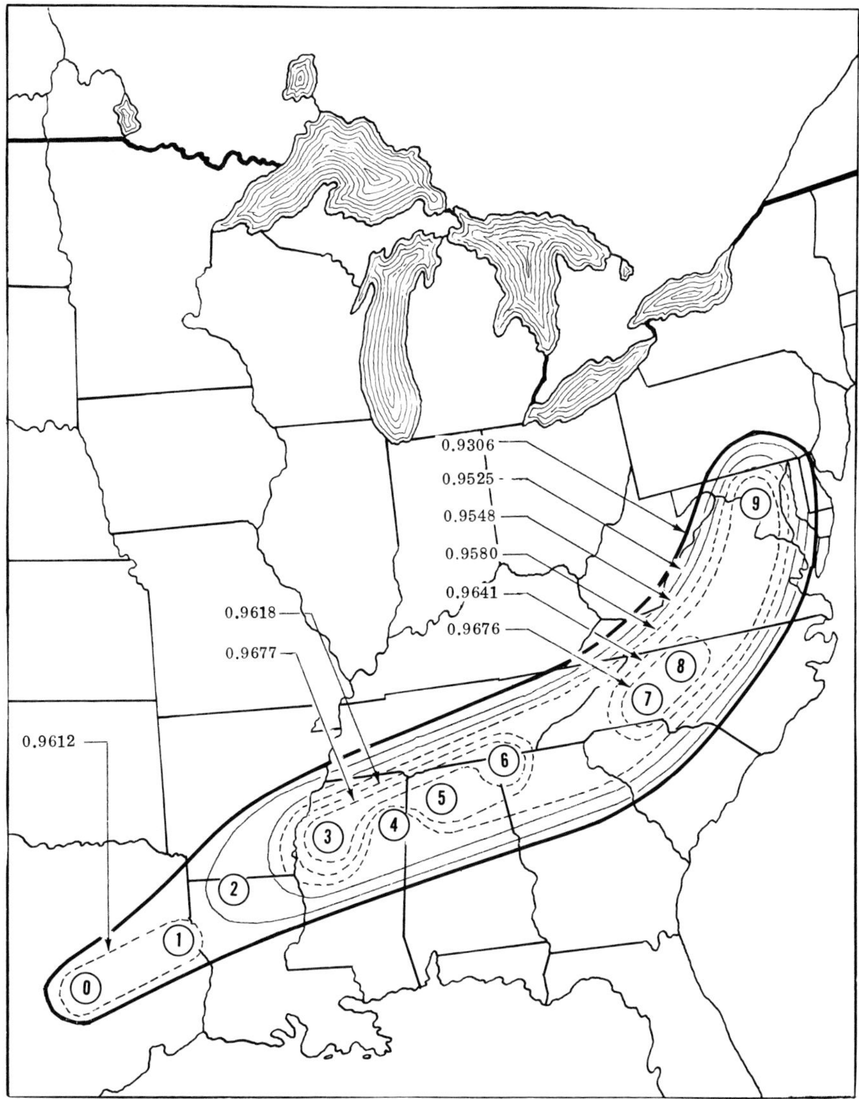

FIG. 3. *Juniperus virginiana* aggregation contours—second year data.

Texas populations were derived but recently (say during the past 5000 years) from the eastern populations.

If the Texas populations were not derived recently from the eastern populations, what other alternatives exist? The most likely is that they are remnants of populations of the now largely Rocky Mountain species, *Juniperus scopulorum.* The latter is very closely related to *J. virginiana* and is distinguished from that taxon with considerable difficulty (Fassett, 1945). The two species possess nearly identical volatile components (Von Rudloff and Couchman, 1964; Vinutha and Von Rudloff, 1968), although detailed quantitative study of their distribution in natural populations of *J. scopulorum* has not been made. In fact, some authors consider the two taxa to be but regional varieties of the same species (Fassett, 1945). A wide range of workers (Wakefield and Jones, 1968) have suggested that the Rocky Mountain conifers and especially *J. scopulorum* (Wells, 1970) expanded their ranges considerably to the southeast during or following the most recent glacial period and that within the last 10,000 years such forests have extended into the central Texas region. Thus, the Texas populations of *J. virginiana* sampled in our transect may have been derived from *J. scopulorum* either directly or as a result of relatively recent hybridization with this species and preexisting populations of *J. virginiana* following the most recent glacial period.

In connection with the above, it is interesting to note that Van Haverbeke (1968), in a detailed study of the northwesternmost populations of *J. virginiana* (Fig. 1), has suggested that this species and *J. scopulorum* hybridize allopatrically over a broad region of the Missouri River Basin. As an alternative explanation for the considerable morphological variation found in this region, however, he suggested that *J. virginiana* arose out of *J. scopulorum* in relatively recent times, the former having extended into the southeastern United States acquiring its subtle distinctions as it migrated into the lower, more mesic habitats of this region.

Without detailed populational data for *J. scopulorum* it is not possible to make a strong case for any of these alternatives. However, present data from *J. virginiana* populations, which show a distinct step-cline from the Appalachian region to eastern Texas, with regional groups of populations entering the "expanding aggregation" together, joining those in the Appalachian area (Fig. 3), suggest that populations 2 through 6 are derived out of the northeast, presumably as a result of gradual dispersal and adaptive accommodation in the local regions along the transect. That is, these results suggest that the clinal pattern was achieved through relatively gradual migration and *in situ* adaptation by drawing repetitively upon the large, relatively homogeneous gene pool of the Appalachian populations.

Additional investigations into the populational structure of both *J.*

scopulorum and *J. virginiana* seem highly desirable, especially in the Missouri River Basin where allopatric introgression reportedly occurs. If, as Van Haverbeke (1968) suggests, the populations in this region are a result of hybridization between these two taxa, then at least some of the populations ought to resemble the Texas populations, supporting the hybridization model for the latter populations suggested above. Whatever the results, however, it seems evident that some of the more difficult systematic problems within the *J. virginiana* complex might yield to the combined efforts of systematist, chemist, and computer analyst.

Summary

Populations of *Juniperus virginiana* sampled at approximately 150-mile intervals along a 1500-mile transect from northeastern Texas to Washington, D. C., in two successive years were analyzed to determine their phenotypic structure. The terpenoid patterns from individual foliage samples of ten plants representing each population in the first year's study were obtained by gas–liquid chromatography. Cluster analysis of the latter indicated a statistically significant clinal pattern of differentiation in which the most mutually similar populations appear in the Appalachian region, and progressively more divergent populations were found in more distant regions as measured along the transect toward the southwestern portion of the United States.

These populations were resampled a year after the first study, together with one additional population, extending the transect about 150 miles deeper into Texas. The number of trees sampled at each population was doubled. These new data were processed in the same manner as the original data. The statistically significant portion of the clustering pattern, indicating clinal differentiation, was reconfirmed. However, the larger sample size provided better resolution, apparent through the emergence of three distinct steps in the cline.

These data permitted several new phyletic interpretations as to the nature of the cline and the origin of the more disjoint Texas populations.

References

Anderson, E. 1953. *Biol. Rev. Cambridge Phil. Soc.* **28**:280.

Barber, H. N., and W. D. Jackson. 1957. *Nature* (*London*) **179**:1267.

Davis, P. H., and V. H. Heywood. 1963. "Principles of Angiosperm Taxonomy." Oliver & Boyd, Edinburgh.

Fassett, N. C. 1945. *Bull. Torrey Bot. Club* **72**:480.

Flake, R. H., and B. L. Turner. 1968. *J. Theor. Biol.* **20**:260.

Flake, R. H., E. Von Rudloff, and B. L. Turner. 1969. *Proc. Nat. Acad. Sci. U.S.* **64**:487.

Grant, V. 1958. *Cold Spring Harbor Symp. Quant. Biol.* **23**:337.
Hall, M. T. 1952. *Ann. Mo. Bot. Gard.* **39**:1.
Hall, M. T. 1955. *Ann. Mo. Bot. Gard.* **42**:171.
Hall, M. T., and C. J. Carr. 1962. *Butler Univ. Bot. Stud.* **14**:21.
Irving, R., and B. L. Turner. 1973. *Taxon* (in press).
Little, E. L., Jr. 1971. *U.S. Dep. Agr., Misc. Publ.* **1146**.
Mayr, E. 1965. *Syst. Zool.* **14**:73.
Sokal, R. R., and P. H. A. Sneath. 1963. "Principles of Numerical Taxonomy." Freeman, San Francisco, California.
Turner, B. L. 1970. *In* "Phytochemical Phylogeny" (J. B. Harborne, ed.), pp. 187–205. Academic Press, New York.
Van Haverbeke, D. F. 1968. *Nebr. Univ. Stud., N.S.* p. 38.
Vinutha, A. R., and E. Von Rudloff. 1968. *Can. J. Chem.* **46**:3743.
Von Rudloff, E. 1968. *Can. J. Chem.* **46**:679.
Von Rudloff, E., and F. M. Counchman. 1964. *Can. J. Chem.* **42**:1890.
Van Rudloff, E., R. S. Irving, and B. L. Turner. 1967. *Amer. J. Bot.* **54**:660.
Wakefield, D., Jr., and J. K. Jones. 1968. "Pleistocene and Recent Environments of the Central Great Plains," Spec. Pub. No. 3. Dep. of Geol., Univ. of Kansas, Lawrence, Kansas.
Wells, P. V. 1970. *Science* **167**:1574.
Zadeh, L. A. 1965. *Inform. Control* **8**:338.

AUTHOR INDEX

Numbers in italics refer to the pages on which the complete references are listed.

D

E

F

G

Q

R

S

SUBJECT INDEX

D

E

J

K

L

M

N

T

U

V

X

Y

Z